CHEMISTRY AND PHYSICS

OF CARBON

Volume 16

CHEMISTRY AND PHYSICS
OF CARBON

A SERIES OF ADVANCES

Edited by

Philip L. Walker, Jr. and Peter A. Thrower

DEPARTMENT OF MATERIALS SCIENCE AND ENGINEERING
THE PENNSYLVANIA STATE UNIVERSITY
UNIVERSITY PARK, PENNSYLVANIA

Volume 16

CRC Press
Taylor & Francis Group
Boca Raton London New York

CRC Press is an imprint of the
Taylor & Francis Group, an **informa** business

First published 1981 by Marcel Dekker, Inc.

Published 2019 by CRC Press
Taylor & Francis Group
6000 Broken Sound Parkway NW, Suite 300
Boca Raton, FL 33487-2742

ISBN 13: 978-0-367-45202-5 (pbk)
ISBN 13: 978-0-8247-6991-8 (hbk)

**Visit the Taylor & Francis Web site at
http://www.taylorandfrancis.com**

**and the CRC Press Web site at
http://www.crcpress.com**

The Library of Congress Cataloged the
First Issue of This Title as Follows:

Chemistry and physics of carbon, v. 1-
    London, E. Arnold; New York, M. Dekker, 1965-

        v. illus.   24 cm.

    Editor: v. 1- P. L. Walker

    1. Carbon.   I. Walker, Philip L., 1928- ed.

QD181.C1C44   546.681

Library of Congress   1   66-58302
ISBN 0-8247-6991-0

PREFACE

In Volume 4 of this series there was a chapter by Walker, Shelef, and
Anderson on "Catalysis of Carbon Gasification." It was prompted pri-
marily by concern about the gasification of moderator graphite in gas-
cooled nuclear reactors. The chapter summarized the literature up
until 1967 and emphasized studies on the use of magnetic measurements
to follow the chemical state of iron during catalysis of the $C-CO_2$
reaction.

Interest in catalysis of the gas reaction of carbon has now
reached a very high level because of the coming importance of synthetic fuels from coal via the route of coal gasification. Dr. McKee
has prepared a most important and comprehensive chapter for this
volume on "The Catalyzed Gasification Reactions of Carbon." He considers all four of the important gas-carbon reactions, that is, reactions of carbon with O_2, CO_2, H_2, and steam. He also integrates what
we have learned from reactions involving such variable forms of carbon
as natural graphite crystals and coal chars.

In this series, we have had a number of chapters written by
experts concerned with electronic properties of carbon, that is,
"Electronic Transport in Pyrolytic Graphite and Boron Alloys by Pyrolytic Graphite" in Volume 2, "Electronic Properties of Doped Graphite"
and "Positive and Negative Magnetoresistances in Carbons" in Volume 7,
"The Electronic Properties of Graphite" in Volume 8, and "Thermoelectric and Thermomagnetic Effects in Graphite" in Volume 12. This
series of chapters is now culminated by Dr. Spain's chapter in this
volume on "Electronic Transport Properties of Graphite, Carbons, and

Related Materials." This chapter is directed to the nonspecialist. It will be particularly useful to chemists and engineers because an extensive bibliography has been given. Dr. Spain emphasizes the electronic properties of important forms of carbon. He relates, in a methodical way, the preparation of carbon materials, relation of preparation to structure, and then relation of structure to electronic properties.

Philip L. Walker, Jr.
Peter A. Thrower

CONTRIBUTORS TO VOLUME 16

D. W. McKee, Physical Chemistry Laboratory, Corporate Research and
Development Center, General Electric Company, Schenectady, New York

Ian L. Spain, Laboratory for High Pressure Science and Engineering
Materials Program, Department of Chemical and Nuclear Engineering,
University of Maryland, College Park, Maryland
    Current affiliation: Department of Physics, Colorado State
University, Fort Collins, Colorado

CONTENTS OF VOLUME 16

CONTENTS OF OTHER VOLUMES

1

THE CATALYZED GASIFICATION REACTIONS OF CARBON

D. W. McKee

Corporate Research and Development Center
General Electric Company
Schenectady, New York

I. INTRODUCTION

The combustion of carbon is certainly one of the most ancient chemical reactions utilized by mankind, and it is obviously true today
that a major portion of the world's energy is generated by the gasification reactions of carbonaceous materials. The accelerating
effects of impurities on such processes have also been known about
for a long period of time. However, until fairly recently the catalytic phenomena were regarded as mysterious and not amenable to ra-
tional interpretation. Most natural carbonaceous materials have
complex and extremely variable structures and they usually contain
appreciable amounts of adventitious inorganic impurities that generally increase, but occasionally decrease, the reactivity of the carbon. In these circumstances it is not surprising that the extensive
literature on carbon gasification catalysis is confusing and often
contradictory. Only in recent years, with the advent of new experimental techniques and the careful study of the behavior of highly
purified and crystalline carbons that contain catalytic additives
in well-defined chemical states, has some degree of understanding
been brought to this complex and important field. Although many
problems of interpretation remain, it is now possible to speculate
rationally on the mechanisms of the catalytic behavior for a wide
range of materials in gaseous environments of significance.

The major factors that influence the rates of the reactions of
carbon with oxygen, carbon dioxide, water vapor, and hydrogen are

1. The concentration of active sites on the carbon surface
2. The crystallinity and structure of the carbon
3. The presence of inorganic impurities
4. The diffusion of reactive gases to the active sites

This review will be mainly concerned with the third factor, although
to a large extent they are all interdependent, for the presence of
impurities can directly influence the concentration of active sites
at which the gasification reactions occur. Equally influential can
be the drastic changes possible in the pore structure, so that dif-

fusional processes can affect the observed values of the kinetic parameters. Also, in some cases, the presence of impurities can bring
about at high temperatures a change in the crystallinity of the carbon. As the purity of industrial carbons is seldom determined and
the pore structure is extremely variable, it is usually unproductive
to attempt to compare the kinetics of gasification of carbons from
different sources. For this reason the emphasis in this review will
be on mechanisms and techniques rather than on the relative ranking
of catalytic activities.

   Comprehensive reviews are available covering the literature on
carbon gasification up to 1966 [1-4]; therefore, no attempt will be
made to survey this earlier work in detail. In the past few years
interest in the field has significantly increased, specifically as a
result of the current revival of coal gasification processes and
their optimization. Notable advances have also been made in the understanding of the catalytic phenomena, so a review of the past decade of activity appears appropriate at the present time. However, in
view of the very large and proliferating literature, encyclopedic
coverage of published material has not been attempted. Rather the
most important recent advances, in the author's opinion, have been
discussed along with enough background information to make the account
intelligible to the general reader.

II. EXPERIMENTAL TECHNIQUES

During the past decade analyses of conventional kinetic measurements of the effects of naturally occurring or deliberately introduced catalysts on carbon gasification rates have once again prolif-
erated and dominated the literature. As in previous years much of
this effort has been expended on carbonaceous materials of unknown
purity so that the results have only qualitative significance, although they perhaps also have, as in the case of the gasification of
coals and chars, some potential technological importance. Nevertheless, recently there has been an increased concern with the micro-

scopic details of the topographical effects induced by catalytically
active particles on highly purified graphite surfaces and also with
the elementary chemical steps involved in the catalytic processes.
These more fundamental studies are slowly unraveling the complex
mechanisms of the catalytic phenomena.

Comparisons of the effects produced by different catalysts have
generally been made, as in the past, by weight-loss measurements,
sometimes combined with analyses of the gaseous reaction products.
The determination of ignition temperatures has also often been employed as a comparison method for ranking the activity of different
additives [5-7] as carbon combustion catalysts.

The topographical features and localized behavior of catalytic
particles during carbon and graphite gasification have been studied
in detail using a variety of microscopic techniques. Optical hot-
stage microscopy continued to be frequently utilized to record the
formation of pits and channels on graphite crystal surfaces [5,6,8]
and the great depth of field and increased resolution of the scanning
electron microscope has been used to photograph the localized pitting
produced in glassy carbons in the vicinity of embedded catalyst particles [9]. Conventional electron diffraction and electron microscopy have been found useful [10] in studying the behavior of silver
particles during the oxidation of single-crystal graphite, and in
recording the development of microporosity during the gasification of
carbons containing metallic impurities [11].

A new technique that has proved extremely useful for examining
the topographical and dynamic effects produced on crystalline graphite by catalytically active particles during gasification reactions
is Controlled Atmosphere Electron Microscopy (CAEM), developed at
AERE, Harwell, U. K., by Baker, Harris, and their associates [12-14].
This transmission microscopy technique enables the details of catalytic gasification to be studied in situ and at very high magnifica-
tion. In a typical arrangement [12] the modified gas reaction stage
A (JEOLCO JEM AGI) and cell B, shown schematically in Figure 1, are
inserted into an electron microscope (JEM 7A) in place of a conventional specimen holder. The cell B seats on the Neoprene O rings C

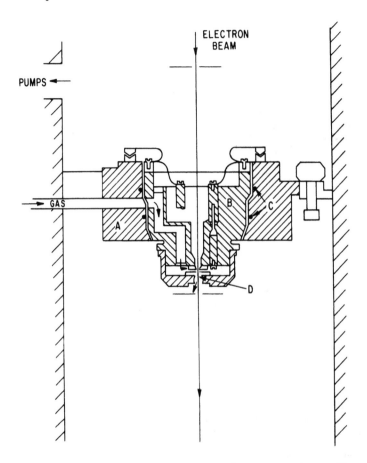

Figure 1. Schematic arrangement of gas reaction cell and stage as
mounted in electron beam for Controlled Atmosphere Electron Microscopy (CAEM) [12]. Copyright 1972 by the Institute of Physics.

to form an annular channel through which the reactant gas passes via
a duct in the stage. The graphite specimen, in the form of a thin
flake, is mounted above the hole D (300 μm in diameter) in a platinum
heater ribbon supported on a mica ring. A Pt/Pt-Rh thermocouple is
spot-welded to a point close to the hole. A direct current passed
through the heater ribbon heats the specimen to the desired reaction
temperature. The specimen chamber, shown by the area enclosed by the
broken line in Figure 2, is evacuated to better than 10^{-5} torr (1.3
x 10^{-3} Pa) by means of an auxiliary vacuum system. After passage

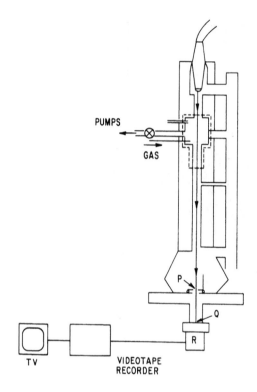

Figure 2. Schematic diagram of the Controlled Atmosphere Electron
Microscope (CAEM) [12]. Copyright 1972 by the Institute of Physics.

through the graphite specimen, the central area of the electron beam
passes through a 20 mm-diameter hole in the center of the fluorescent
screen P and strikes a secondary screen Q with a transmission phos-
phor, the underside of which is viewed with a high-sensitivity camera
R. The output of the camera can be monitored and recorded continu-
ously on videotape or by cinematography of the video display. Under
favorable conditions the limit of resolution of the microscope is approximately 30 Å. This technique has been successfully utilized to
record the details of channeling and pit formation by metallic catalyst particles during the gasification of graphite single crystals by
oxidizing gases at temperatures up to 1200°C and, also, to derive
quantitative estimates of the kinetic parameters of the catalyzed reactions [15-19]. The results of these studies will be discussed at
some length in this review.

Recently [10], x-ray photoelectron spectroscopy has been employed in an attempt to establish the chemical state of silver particles during the Ag-catalyzed oxidation of graphite. This is an emerging technique, which is likely to be utilized increasingly in the interpretation of the mechanism of catalysis.

Thermal analysis, especially the combined application of thermogravimetry and differential thermal analysis, although an old technique, is now proving useful in interpreting the individual steps that are involved in catalyzed gasification reactions [5,6,20,21]. By measuring the rate of evolution of the gaseous oxidation products when heating doped carbons in oxygen at a linearly increasing temperature in a thermobalance, the activities of a range of catalytic additives can be compared and an estimate of apparent activation energies can be obtained [22].

Another technique that has proved useful in identifying the active chemical species involved in the catalytic process is the measurement of magnetic susceptibility. Thus, changes in the paramagnetic component of the magnetic susceptibility during the oxidation of $CuSO_4$-impregnated charcoal [7] have lent support to an oxidation-reduction cycle mechanism involving Cu/CuO as the active catalytic entity.

Some evidence for the intercalation of alkali metals between the basal planes of graphite during catalyzed gasification has been found by means of electron spin resonance (ESR) [23,24]. ESR has also been used to identify the oxidation states of vanadium-oxide catalysts during carbon oxidation [25].

III.   CATALYZED GASIFICATION OF CARBON BY MOLECULAR OXYGEN

A.   Introduction

The exothermic reactions of carbon with molecular oxygen [26],

$$C(graphite) + O_2(g) = CO_2(g) \quad \Delta H°_{298.15 \text{ K}} = -94.054 \text{ kcal mol}^{-1}$$

$$\Delta G°_{1000 \text{ K}} = -94.628 \text{ kcal mol}^{-1}$$

$$C(graphite) + \frac{1}{2} O_2(g) = CO(g) \quad \Delta H°_{298.15\ K} = -26.42 \text{ kcal mol}^{-1}$$

$$\Delta G°_{1000\ K} = -47.859 \text{ kcal mol}^{-1}$$

are favored thermodynamically at all temperatures up to 4000 K, the CO/CO_2 ratio in the primary products increasing with an increasing reaction temperature. At 800°C, a typical combustion temperature in the absence of a catalyst, the rate of the $C-O_2$ reaction is about five orders of magnitude faster than the $C-H_2O$ and $C-CO_2$ reactions and about eight orders of magnitude faster than the $C-H_2$ reaction [2]. The literature on the catalyzed $C-O_2$ reaction exceeds that on all the other catalyzed gasification reactions combined, but a complete interpretation of the catalytic phenomena remains elusive, in spite of some recent advances in understanding. Materials of a very wide range have been found to be active catalysts for the $C-O_2$ reaction and recent work on these will be discussed separately, according to their chemical types.

The presence of catalysts generally results in a decrease in the CO/CO_2 ratio in the products at a given temperature, whereas inhibitors of carbon oxidation have the opposite effect. Interpretation of these changes in product ratio is complicated by the fact that many carbon-oxidation catalysts, for example CuO, NiO, Ag, and Pt are also excellent catalysts for CO oxidation to CO_2, whereas compounds such as $POCl_3$ inhibit both the oxidation of solid carbon and that of gaseous CO. Hence changes in the CO/CO_2 ratio upon addition of catalysts may be a result of secondary processes rather than a consequence of elementary steps occurring at the carbon surface.

It may be noted in passing that although the reaction between carbon and N_2O is generally catalyzed by the same materials as the carbon-O_2 reaction [27,28], attempts to catalyze the carbon-SO_2 reaction have not been particularly successful [29]. Because of the lack of definitive experimental data these oxidation reactions will not be further discussed.

B. Alkali-Metal Oxides and Salts

Alkali-metal salts are among the oldest known additives which markedly increase the rate of oxidation of carbonaceous materials, and
there have been many practical applications of their catalytic
effects. In general, the alkali carbonates are more active than
other salts such as sulfates or halides. It has been known for some
time that the reactivity of a coke toward air can be increased markedly by the addition of sodium or potassium carbonate to the precursor coal before carbonization [30]. Thus, Patrick and Shaw [31]
found that increasing amounts of Na_2CO_3, in the range 0.5-5 wt %,
when added to a coal before carbonization caused a progressive decrease in the ignition temperature of the resulting coke in air.
However, addition of more than 2 wt % of the salt resulted in a saturation effect with only a small further reduction in ignition temperature with further increases in salt concentration. The alkali carbonate proved to be an even more effective catalyst when added to
the cokes after carbonization, presumably because when added to the
coal, some of the salt was reduced to the metal and vaporized during
carbonization. However, a considerable proportion of the sodium initially added was found to be retained by the coke in a strongly bonded
form. The authors suggest that this is evidence for an electronic
mechanism for the catalytic effect. Moreover, the presence of the
alkali carbonate was also found to affect the physical properties of
the coals. Thus, addition of 5 wt % Na_2CO_3 greatly reduced the flu-
idity of the coal during carbonization and increased the porosity of
the resulting coke. These physical changes render a definitive interpretation of the catalytic effect difficult.

It has been assumed that studies of catalytic gasification reactions of coals and chars were likely to be complicated by the presence in these natural materials of ubiquitous impurities and minerals
that may themselves exert a strong catalytic effect. However, Tomita
et al. [32] studied the reactivities of various char-mineral mixtures
in air and concluded that the most common minerals found in coal have
little catalytic effect on the coal-air reactivity. Nevertheless, Ca

and Mg impurities can catalyze the carbon-O_2 reaction and, as noted in Section VI.B, iron minerals, such as pyrite and siderite, can be active catalysts in the hydrogasification of coal and chars.

The catalytic effect of sodium carbonate on the coal-air reaction is utilized in the Kellogg Molten Salt Coal Gasification Process for the production of low BTU substitute natural gas [33,34]. In this process, coal is added to a bath of molten carbonate at 925°C and compressed air is injected at 260-400 psia to partially burn the coal in order to yield a fuel gas (essentially a mixture of N_2, CO, CO_2, and H_2) with a heating value of 100-150 BTU/SCF at a conversion efficiency of about 90% [35].

Harker et al. [24] showed that potassium impurity, introduced either electrochemically or from the vapor phase, had a marked effect in accelerating the gasification of neutron-irradiated graphite by oxygen. The higher combustion rate appeared to be due almost entirely to a higher rate of formation of CO_2, the rate of formation of CO remaining essentially unchanged on addition of the alkali metal. An electronic mechanism was invoked to explain the catalytic effect, the ionization potential of the graphite aromatic layers being assumed to be altered by K atoms at localized sites or lattice imperfections.

In an attempt to unravel the mechanism of the catalytic process, McKee and Chatterji [21] studied the effects of alkali-metal carbonates and oxides on the oxidation of graphite by oxygen by means of simultaneous TGA-DTA and by hot-stage microscopy. As shown in Figure 3, which illustrates TGA heating curves for graphite and graphite-Na_2CO_3 mixtures, the presence of the alkali carbonate resulted in a marked decrease in the temperature required for a given amount of gasification, and in separate experiments it was found that catalytic oxidation at temperatures in the vicinity of 600°C was accompanied by the presence of mobile liquid droplets and channeling of the graphite basal plane. Addition of the alkali-metal monoxides, for example Na_2O, to graphite powder also resulted in a very rapid acceleration in the rate of gasification in oxygen, the reaction being initiated at temperatures in the range 450-500°C. In this temperature range

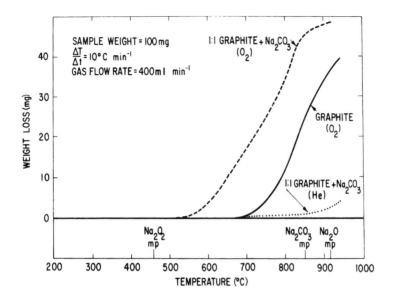

Figure 3.   Thermogravimetric (TGA) curves for graphite powder and
1 to 1 graphite-Na_2CO_3 mixtures on heating in He and O_2 [21].

Na_2O rapidly reacts with O_2 to form the peroxide, Na_2O_2, as shown by
the TGA-DTA curves for Na_2O heated in O_2 in Figure 4. As peroxides
of the alkali metals react readily with carbon at low temperatures,
the authors conclude that the catalytic process involves an oxidation-
reduction cycle with the intermediate formation of peroxide or higher
oxide.   Such a mechanism is consistent with thermodynamic reasoning,
as will be discussed in Section III.A.

C. Alkaline Earth Oxides and Salts

Although the strong catalytic effect of barium salts, especially
the peroxide BaO_2, on carbon oxidation rates is well known [36], the
catalytic behavior of other compounds of the group IIA metals has
been less studied.
        Calcium salts, in the form of lime or dolomite, are common im-
purities in natural carbonaceous materials such as coal and may have

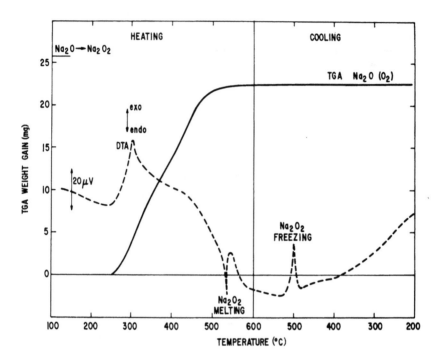

Figure 4. Thermogravimetric (TGA) and differential thermal (DTA) curves for Na_2O heated and cooled in flowing O_2. Gas flow rate = 400 ml min^{-1}. Sample weight = 100 mg. $\Delta T = 10°C$ min^{-1} [21].

an effect on the reactivity of the coal towards oxidizing gases. Thus, Jenkins et al. [37] measured the reactivity of a number of coal chars toward air at 500°C, the chars being prepared by carbonizing coals of widely varying rank at 1000°C under controlled conditions. A correlation between the rank or carbon content of the original coal and the reactivity of the resulting char was obtained, lower-rank coal chars being more reactive than those prepared from high-rank coals. The most reactive chars from low-rank coals contained the highest amounts of Ca and Mg impurities. Chars made from coal that had been washed with acid to remove mineral impurities were significantly less reactive than chars from unwashed coals.

Further evidence of the active catalytic effects of Ca salts has been provided by Cairns et al. [38] who, by means of CAEM combined

with the electron microprobe, identified CaO as the catalytic entity
responsible for the enhanced oxidation rate of impure graphite. Catalytic pitting was also observed when single-crystal graphite was deliberately contaminated with CaO and heated in oxygen in the electron
microscope.

However, in comparison with the oxides of the alkali metals and
even with those of the other alkaline earth metals, the catalytic
activity of CaO in carbon oxidation does not appear to be outstanding. For example, in the course of a recent thermogravimetric investigation by the present author [21,39], it was observed that, unlike the alkali-metal oxides, the oxides of the group IIA metals exhibited only modest catalytic effects when small amounts (1 wt %)
were mixed with powdered graphite and then subsequently heated in
air. Table 1 shows the temperature at which 1 wt % of the carbon
had burnt off when 1 wt % MgO, CaO, SrO, or BaO was initially mixed
with the graphite. Corresponding temperature values for graphite
with 1% added Li_2O and Na_2O are included for comparison purposes.
In the alkaline earth oxide series the relative catalytic effects
were in the order BaO > CaO > MgO > SrO.

With graphite impregnated with aqueous solutions of soluble salts,
followed by freeze drying, the catalytic activity for graphite oxida-

Table 1. Effects of alkali and alkaline earth oxides on
the oxidation of graphite in air

Material	T (°C) for 1% burnoff in flowing air (400 ml min^{-1}) ($\Delta T/\Delta t = 10°C\ min^{-1}$)
Pure graphite powder (G)	763
G + 1% MgO	740
G + 1% CaO	728
G + 1% SrO	750
G + 1% BaO	718
G + 1% Li_2O	568
G + 1% Na_2O	608

Source: From Refs. 21 and 39.

tion was Na > Ba > Sr > Mg > Ca, according to Amariglio and Duval
[36]. As has been demonstrated many times before [27], the degree
of dispersion of the catalyst on the carbon surface is very important
in determining the magnitude of the catalytic effect, and the con-
trasting patterns of catalytic activity observed by different authors
may be due to this factor, among others.

D. Transition Metals and Oxides

The most popular carbon oxidation catalysts selected for detailed
study continue to be the transition metals and their salts and
oxides. In the most recent of a series of investigations into the
reactivities of metal-doped polymer carbons, Marsh and Taylor [40]
examined the catalytic effects of finely dispersed V, Cr, Mn, Fe, and
Co deposited from solution in a polyvinyl acetate that was subsequently carbonized to give a very fine dispersion (3-100 nm diameter)
of metal particles throughout the glassy carbon matrix. On subsequent gasification in oxygen the magnitude of the catalytic effects
for the different metals were in the order Co > V > Fe > Mn > Cr.
The reason for the high activity of Co is not known and has not been
noted by others in the field, who have generally reported V to be the
most active of this group. The degree of metal dispersion may be the
important factor in this case. Previously, Marsh and Adair [28]
doped furfuryl alcohol with a series of metal nitrate solutions. On
polymerization and subsequent carbonization, the metal was again dispersed as very small aggregates. In this case the relative catalytic
activities of the doped carbons toward oxygen were in the order
Ni > Co > Cu > Ag > Fe > Ca (inactive), at an impurity level of 400
ppm.

Using the lowering of the ignition temperature of pure graphite
powder in dry oxygen as a criterion for catalytic activity, the author [6] obtained the order of activity: Pb > V > Mn > Co > Cr,
Fe > Ni, as shown in Table 2.

A detailed study of the catalytic effect of vanadium and its

Table 2. Effect of catalysts on the ignition temperature
of graphite powder in O_2

Catalyst (acetate or oxide)	w/o as metal	Ignition T (°C) in O_2
Pb	0.15	382
V	0.20	490
Mn	0.45	523
Co	0.33	525
Cr	0.95	540
Cu	0.20	570
Mo	0.15	572
Ag	0.16	585
Cd	0.21	590
Fe	0.13	593
Pt	0.03	602
Ni	0.45	613
Ir	0.40	638
Rh	0.20	622
Ru	0.30	640
Pd	0.30	659
Ce	0.72	692
Zn	50.00	700
W	0.02	718
Hg	0.10	720
Sn	0.10	738
Pure graphite	—	740

Source: From Ref. 6.

oxide has been made by Baker et al. [17], using the CAEM technique.
Both vanadium metal (applied by vacuum evaporation) and V_2O_5 (applied
by atomized-spray deposition) were extremely active catalysts for the
oxidation of single-crystal graphite at temperatures around 550°C.
Liquid-droplet formation and catalytic channeling accompanied the
gasification process, and there was visual evidence that V_2O_5 pene-
trated between the graphite layers. After the reaction an intermediate oxide V_6O_{13} was identified on the graphite surface by electron
diffraction. The transient formation of V_6O_{13} was also confirmed in
the thermogravimetric studies by the present author of the V_2O_5-

catalyzed graphite-oxygen reaction [6]. As shown in Figure 5, a mixture of equal weights of V_2O_5 and powdered graphite, on heating together in the thermobalance in an inert atmosphere, reacted at 675°C
with partial reduction of the V_2O_5 to the lower oxide V_6O_{13}. An
oxidation-reduction cycle that accounts for the observed catalytic
effect of vanadium was proposed (Section III). CAEM studies of the
catalytic effects of molybdenum and MoO_3 on graphite single crystals
have also been carried out [16]. In this case also pitting and
channeling behavior was observed in the temperature range 600-800°C.
Exfoliation of the graphite sample was attributed to penetration of
MoO_3 between the layer planes of the graphite. There was, however,
no direct evidence for the formation of a graphite-MoO_3 intercalation
compound. Chromium, probably in the form Cr_2O_3, has been reported to
be an active catalyst for graphite oxidation [6,27], but its level of
activity was not outstanding.

The catalytic effects of iron have also been investigated by
CAEM [41] and it has been demonstrated that iron impurity added to
neutron-irradiated, moderator graphite leads to a considerable in-

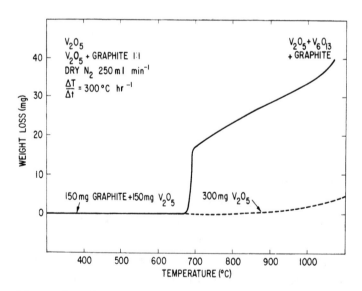

Figure 5. Thermogram for a 1:1 graphite-V_2O_5 mixture on heating
in dry N_2 [6].

crease in the reactivity of the material toward molecular oxygen
[24]. Iron particles cause both pitting and channeling in the graphite basal plane surfaces [6]. Adair et al. [9] have examined the
pitting morphology produced by Fe, Ni, and other metals on oxidation
of polyfurfuryl alcohol carbons containing fine dispersions of the
metal particles. Localized pitting and preferential gasification in
the vicinity of metallic inclusions produced honeycomb structures
with interconnected pores and crevices.

E. Noble Metals

Silver is known to be a very active catalyst for carbon gasification and has been the subject of a number of investigations.
Turkdogan and Vinters [42] reported that 1.7 wt % Ag added to electrode graphite increased the rate of oxidation in air at 600°C by
almost 3 orders of magnitude, whereas 1 wt % Ni had very little
effect.

    A detailed study of the effects of Ag in the graphite–oxygen re-
action has been made by the CAEM method [43]. Somewhat surprisingly
it was found that although Ag acted as a powerful catalyst for the
oxidation reaction at temperatures below 900°C, at higher temperatures it appeared to inhibit the reaction. The effect was rationalized in terms of the "compensation effect," whereby the catalyzed
reaction with a smaller activation energy than the uncatalyzed reaction would be more rapid than the latter at temperatures below the
isokinetic temperature and less rapid than the uncatalyzed reaction
at higher temperatures. The channeling and pitting behavior of Ag
particles showed some interesting features and will be discussed further below (Section III.H). The behavior of Ag particles during the
oxidation of single-crystal graphite at 477°C has also been recently
studied by transmission electron diffraction and microscopy [10].
Mobile clusters of Ag particles formed channels on the graphite surface when heated in oxygen, the channels originating at steps on the
graphite surface. X-ray photoelectron spectra of the graphite coated

with Ag both before and after oxidation showed no significant differences and no evidence for the formation of bulk oxide of the
metal; the active catalytic entity was therefore probably the metal
itself. This conclusion was also reached by the present author on
the basis of the results of a thermogravimetric study of the behavior
of graphite-silver acetate mixtures on heating in oxygen [6]. No
evidence for the formation of a separate Ag_2O phase was found, silver
acetate decomposing completely to the metal at temperatures in the
range 250-275°C.

In the same study [6] gold was observed to be a less active catalyst than silver; however, there have been earlier reports of a high
catalytic activity for gold [44], and it is known that catalytic
channeling of graphite surfaces can be induced by finely divided
gold particles [41,45].

Although the Pt group metals are not the most active catalysts
for the oxidation of carbon, their behavior in this connection is of
considerable practical importance, as an oxidation step is usually
employed to reactivate a carbon-contaminated, supported noble-metal
catalyst. Probably for this reason there have been a number of investigations into the behavior of this group of metals. The author
[6] observed that Pt, Ir, Rh, Ru, and Pd were fairly active catalysts
for graphite oxidation and produced substantial decreases in the ignition temperature of finely powdered graphite (Table 2). The most
active metal of the group was Pt and the least active Pd. Irregular
etch pits were observed to form on graphite crystal surfaces during
oxidation in the presence of finely dispersed noble-metal particles.

Fryer [46] found some indication that Pd particles behaved as
liquid droplets during the oxidation of graphite flakes at 500°C in
air. Liquidlike behavior may be expected at temperatures exceeding
the Tamman temperature ($0.52T_m$ K = 676°C for Pd), indicating that appreciable local heating accompanied the exothermic oxidation reaction
in the vicinity of a catalyst particle. Channels were formed by the
moving Pd particles, or droplets, which, as with Mo, were sometimes
observed to penetrate between the graphite layers.

The behavior of Pt and Pd particles on the surface of graphite

single crystals during oxidation has been studied in detail by Baker
et al. [18], using the CAEM technique combined with video cinematography. At 500°C and below, particles of both metals in the 2 to 5
nm-diameter size range produced pits in the graphite basal plane
(Figure 6). The pits subsequently expanded by uncatalyzed oxidation
of the edges to form hexagonal holes. At higher temperatures (735-
850°C) the metal particles became mobile and formed channels in the
graphite surface that were essentially straight for the smaller particles and usually oriented parallel to the < 11$\overline{2}$0 > directions (Figure
7). Particles larger than 100 nm in diameter formed irregular channels with hexagonal facets at the graphite-catalyst interface, the

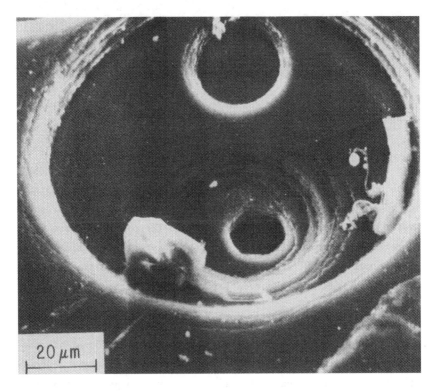

Figure 6. Pit produced by Pt catalyst on graphite crystal surface
on heating in 5 torr O_2 at 500°C [18].

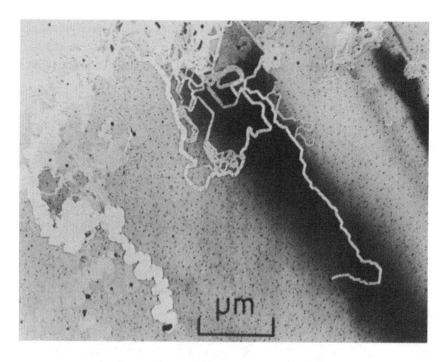

Figure 7. Catalytic channeling produced on a graphite crystal surface by Pt particles at 750°C in 5 torr O_2 [18].

facets being oriented parallel to the < $11\bar{2}0$ > directions. At 850°C
all the moving particles were spherical in form and exhibited liquidlike properties such as coalescence and fragmentation. The presence
of water vapor in the O_2 atmosphere reduced the oxidation rate of
the graphite but had little effect on the apparent activation energy
($\sim$84 kcal mol^{-1} for Pt and $\sim$105 kcal mol^{-1} for Pd). A similar effect
of water vapor has been observed for the Cu-catalyzed oxidation of
graphite [5].

    CAEM has also been used recently to investigate the behavior of
dispersed alloy particles during the catalytic oxidation of single-
crystal graphite [47]. Thus the channeling characteristics of Pt-Ir
particles resembled those of Ir rather than Pt, suggesting that the
surface of the particles had become enriched in Ir in the oxidizing
atmosphere.

In the course of a detailed study of the kinetics of sintering of Pt crystallites on carbon substrates, Chu and Ruckenstein [48] found that Pt particle mobility and the rate of crystallite growth at 500°C were much greater in an atmosphere containing O_2 than in a vacuum. The rate of sintering increased as the oxygen pressure was raised from 10^{-4} to 10^{-2} torr (0.013-1.3 Pa). Not unexpectedly the sintering process in O_2 was accompanied by catalytic channeling and gasification of the carbon surface. Of particular interest in this work was the observation that Pt crystallites dispersed on carbon reacted with O_2 or water vapor at 500°C to give a surface platinum-oxide phase which was identified by electron diffraction and dark-field electron microscopy. The operation of an oxidation-reduction cycle involving the sequential formation and reduction of platinum oxide on the carbon thus appears to be a feasible mechanism for the catalytic effect of Pt and probably also for the other noble metals. The progressive growth of Pt crystallites during catalytic channeling of a graphite substrate at 1100 K in an O_2 pressure of only 10^{-6} torr $(1.3 \times 10^{-4}$ Pa) has recently been observed by transmission electron microscopy [49].

F. Miscellaneous Catalysts

In addition to those mentioned above, a wide variety of other metals, oxides, and salts have demonstrated significant catalytic activity in carbon oxidation by molecular oxygen. Of particular interest are copper and lead which are among the most active catalysts known for this reaction. The Cu-catalyzed oxidation of carbon may play a detrimental role in accelerating the wear of carbon brushes sliding against copper commutators or slip-rings [50]. The catalytic properties of Cu and Pb have been utilized in the oxidative surface treatment of high-strength carbon fibers to improve the interlaminar shear strength of fiber-epoxy composites [51]. Copper compounds have also been added to oven coke to improve its combustion properties [52].

Several studies have been made of the catalytic effects of Cu and its salts in accelerating the oxidation of graphite [5,7,42]. Turkdogan and Vinters [42] found that the addition of 1.1% Cu to electrode graphite granules increased the rate of oxidation in air at 600°C by about 3 orders of magnitude. Similarly, the present author [5] showed that the presence of small amounts of Cu salts had a very marked effect on the rate of oxidation and appreciably lowered the ignition temperature of graphite in oxygen (Table 2). The presence of water vapor reduced the rate of the Cu-catalyzed oxidation but had little effect on the ignition temperature. Hot-stage microscopy revealed that during the catalyzed oxidation at 700-800°C, copper oxide particles in the 1-5 μm size range migrated rapidly on the basal-plane surface of graphite single crystals, cutting irregular channels in random directions. Thermogravimetric measurements suggested that the catalytic effect involved an oxidation-reduction cycle with the localized reduction of CuO in contact with the graphite substrate, followed by reoxidation of the resulting metal to oxide by the ambient oxygen. Similar conclusions were reached by Patrick and Walker [7] who investigated the Cu-catalyzed oxidation of a charcoal by means of gas analysis and magnetic-susceptibility measurements. Cu added as $CuSO_4$ caused a marked reduction in the ignition temperature of the charcoal in O_2. The presence of SO_2 in the gas phase and changes in the magnetic susceptibility indicated that metallic Cu was formed at low temperatures by reaction of the $CuSO_4$ with the carbon substrate. These results provide additional evidence for the occurrence of a reaction sequence wherein the metallic copper is alternately oxidized by gaseous O_2 and the resulting CuO is subsequently reduced to the metallic state by reaction with the carbon, which is simultaneously converted to gaseous oxides of carbon.

Cadmium oxide has been reported to be an active catalyst for the oxidation of graphite [27], although its activity based on ignition-temperature lowering was less than that of Cu or Ag (Table 2) [6].

Varying results have been reported for zinc by different authors. McKee [6] found that graphite powder containing 50 wt % of ZnO had an

ignition temperature in oxygen only 40°C lower than that of pure
graphite (Table 2). Also, alloying the Cu particles with Zn to form
a brass resulted in a marked reduction in the catalytic activity of
the metallic particles when mixed with powdered graphite [5]. These
results are not unexpected, as ZnO is a stable oxide that is not
readily reduced by carbon; however, if the ambient partial pressure
of CO is below 10^{-3} atm, reduction may occur at 700°C or less. Catalytic channeling has been observed with ZnO particles on graphite
single crystals at temperatures around 800°C using the CAEM technique
[41,53]. Baker and Harris [53] estimate an isokinetic temperature of
1097°C for Zn on graphite, so that only below this temperature would
Zn be expected to be active as a carbon-oxidation catalyst, assuming
that the presence of Zn causes a reduction in the activation energy
of the oxidation reaction. Also, Turkdogan and Vinters [42] have
reported enhanced oxidation rates of electrode graphite when impregnated with 0.1% zinc in the temperature range 600–800°C. It is possible that at these temperatures ZnO particles act as sites for the
dissociation of molecular O_2.

Of the series of metals and oxides examined by the present
author [6], lead oxide (from lead acetate) reduced the ignition
temperature of powdered graphite more than any other additive (Table
2). Also, lead acetate crystals evaporated from aqueous solution
onto the basal-plane surface of graphite flakes in the hot-stage
microscope decomposed to oxide between 200 and 300°C. On heating
in oxygen, vigorous motion of the catalyst particles commenced at
500°C with the formation of irregular channels and pits on the graphite surface. The moving particles appeared as liquid droplets and
the growing pits were often lined with dark films, which were probably
metallic lead. PbO is reduced to the metallic state by carbon at
temperatures in the catalytic oxidation range and again the catalysis
process appears to involve a cyclic conversion of PbO to metal and
back to oxide during the carbon gasification reaction. Lead was also
the most active catalyst of the series of metals examined by Magne
and Duval [27], which included Na, Ag, and V.

However, in the course of a recent study of the oxidation behavior of sucrose carbon containing various impurities [54], it was found that Pb increased the apparent activation energy of the reaction and, when present together with vanadium, diminished the catalytic activity of the latter. This requires confirmation in view of the finding by most investigators that lead and lead compounds are very active catalysts for carbon oxidation. A recent study of the Pb-catalyzed oxidation of single-crystal graphite has been made by Harris et al. [55], using the CAEM technique. The channeling behavior of lead oxide particles (again formed from lead acetate) was studied in detail as a function of temperature. At temperatures below 547°C the catalyzed reaction was accompanied by the recession of basal-plane steps at very high rates, as shown in Table 3. Each step receded initially in a direction normal to the step edge, but at a later stage catalytic channels appeared starting from the step edge, and further recession of the step ceased. The authors interpret these observations in terms of an extremely thin film of lead, which initially occupied the face of the step. As the step receded and lengthened the lead film fractured into discrete beadlets of catalyst, which cut channels into the step. At temperatures above 547°C rapid edge recession was not observed but a pattern of rings (probably expanded vacancies decorated by lead) of irregularly shaped particles appeared over the whole graphite surface.

Table 3. Effect of Pb on rates of recession
of graphite basal-plane steps

$T°C$	P_{O_2} (kPa)	Edge recession rate (nm sec^{-1}) with Pb catalyst (V_1)	Uncatalyzed (V_2)	V_1/V_2
495	0.8	18 ± 1	9×10^{-5}	2.2×10^5
530	0.9	49 ± 5	6×10^{-4}	8.5×10^4

Source: From Ref. 55.

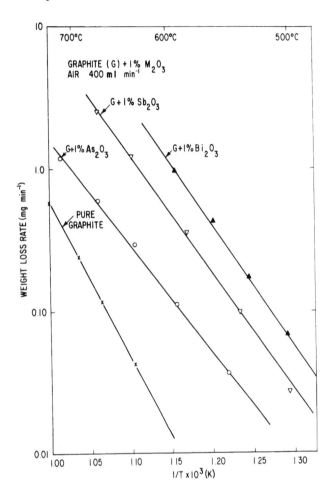

Figure 8.   Catalytic effects of group VB oxides on the oxidation of graphite in air.   Gasification rates vs. 1/T K.   Sample weight = 200 mg [56].

In contrast to P_2O_5, which is an oxidation inhibitor (see Section III.G), the trioxides of the group VB elements As, Sb, and Bi are moderately active catalysts for carbon oxidation, as illustrated by the Arrhenius plots shown in Figure 8 [56].   On addition of 1 wt % of the oxides to graphite powder an increase in oxidation rate and a decrease in apparent activation energy was observed.   At 600°C, 1% Bi_2O_3 increased the oxidation rate of graphite by nearly two orders of magnitude. As_2O_3 was the least active of the triad of oxides, possibly because this oxide is appreciably volatile at temperatures above the melting point (315°C).

Stannous oxide SnO also showed a slight catalytic activity for graphite oxidation [56] but the more stable stannic oxide SnO_2 was inactive.

The interaction of boron and molybdenum catalysts with lattice imperfections and dislocation cores on the basal-plane surfaces of graphite crystals has been demonstrated by Roscoe [57]. In the presence of the moist oxygen used in these experiments, B and Mo probably formed boric and molybdic acid derivatives which were strongly chemisorbed at the exposed carbon atoms at lattice imperfections. Roscoe suggests that in the case of boron, a chemisorbed complex of the form

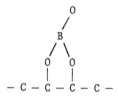

is the active catalytic species. However, although the association of the oxides of B and Mo with lattice imperfections is well known and has been used in the determination of dislocation density in natural graphite crystals [58], the mechanism of the catalytic oxidation effect is not clear. B_2O_3 is a stable oxide that is not likely to be reduced by carbon at temperatures in the carbon oxidation range. The catalytic behavior of boron is, however, influenced in a complicated manner by the presence or absence of water vapor and by the temperature of the carbon-oxidation reaction. Using temperatures in the neighborhood of 800°C, Thomas and Roscoe [59] found that B acted as a catalyst for graphite oxidation in moist oxygen but as an inhibitor in dry oxygen. They concluded that the catalytic effect was due to the formation of volatile boric acids on the graphite surface, whereas in dry oxygen B_2O_3 tended to block the active sites. Subsequent work by Allardice and Walker [60,61] indicated that graphite

doped substitutionally with B had lower gasification rates in the
temperature range 600-650°C than did pure graphite in both moist and
dry oxygen, in spite of the fact that the boron lowered the activation energy of the reaction. It was also observed that the decrease
in activation energy with increasing B content correlated with the
lowering of the Fermi level of graphite on addition of B, thus lending support for the general idea of electron transfer between the
graphite matrix and the B impurity. At the higher temperatures used
by Thomas and Roscoe, removal of B_2O_3 as volatile boric acids from
the graphite surface would be likely, but the observed catalytic
effect is difficult to explain. Woodley [62,63] also studied the
effect of 5% B on the gasification of nuclear graphites in oxygen and
observed a marked inhibitory effect which he attributed to the presence of boric oxide species on the graphite surface. When this surface contamination was removed by leaching in boiling water the oxida
tion rate was increased. It is evident that the influence of tempera
ture on the behavior of B in carbon oxidation reactions would be wort
further study.

G. Inhibition of the C-O_2 Reaction

Although a large variety of metals, salts, and oxides catalyze the
oxidation of carbon, very few additives are known to effectively
inhibit this reaction. In view of the many technical applications
of carbon in which resistance to oxidation would be an advantage,
it is surprising that so little attention has been paid to this area
Additives that have been reported to act as oxidation inhibitors are
mostly compounds of phosphorus and the halogens and, to some extent,
boric acid and its derivatives.

The inhibitory effects of chlorine and its compounds such as
CCl_4 and CCl_2F_2 have been well documented [64], although the situation is complicated by the observation that small concentrations of
halogens may in fact accelerate the gasification of high-purity
graphite in CO_2 [65]. The inhibitory effect at higher halogen

concentrations appears to involve the formation of a stable carbon-
halogen complex at unsaturated sites on the carbon surface [66].

Phosphorus pentoxide is also an effective oxidation inhibitor
and even the vapor of P_2O_5 in the pressure range 10^{-2}-1 torr (1.3-
130 Pa) has a marked retarding effect on the rate of oxidation of
graphite in air at temperatures in the vicinity of 600°C [67]. About
the most powerful inhibitor known is phosphorus oxychloride, whose
interaction with carbon and graphite has been the subject of a number
of investigations. For example, $POCl_3$ vapor has been found to be
effective in retarding the oxidation of nuclear graphites in air at
500°C [68], the ratio of the uninhibited reaction rate to that of the
$POCl_3$-inhibited rate ranging from 11 to 29 at this temperature. The
presence of $POCl_3$ is known to retard the homogeneous reaction between
CO and O_2 and as the CO/CO_2 ratio in the products of carbon oxidation
have been found to be increased in the presence of this additive,
Arthur and Bangham [69] suggested that a similar mechanism may be
involved in the heterogeneous reaction of O_2 at a carbon surface.
The conclusion of Wicke [70] that $POCl_3$ presents a physical barrier
to oxidation by forming a strongly adsorbed layer over the entire
carbon surface cannot be the complete explanation as the effective
concentrations of $POCl_3$ required to markedly reduce the oxidation
rate are generally very much less than that necessary to cover the
carbon surface with a complete monolayer.

Recently optical studies by the author on the effect of $POCl_3$
on etch pit growth on graphite single crystals during oxidation [71]
have indicated that this additive is strongly adsorbed at dislocation
cores and on the edges of preformed etch pits in the basal plane.
Small amounts of $POCl_3$ remain strongly adsorbed on the graphite surface for long periods of time at temperatures in the range 700-800°C.
At these temperatures the rate of growth of etch pits on the graphite
basal plane was found to be dramatically decreased following treatment with $POCl_3$, as shown in Figure 9. Estimates of the activation
energy for oxidation in the $< 10\bar{1}0 >$ directions for the $POCl_3$-treated
graphite ranged from 100 to 150 kcal mol^{-1}, compared with the corresponding value of 66 kcal mol^{-1} for the uninhibited oxidation rate as

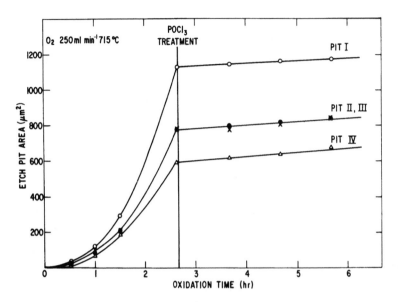

Figure 9. Effect of POCl₃ on the rate of growth of individual etch pits on crystal graphite at 715°C in dry O₂ [71].

reported by Thomas [72]. The inhibitory effect of $POCl_3$ was further enhanced by pretreatment of the graphite with a mild oxidant such as sodium hypochlorite solution. This pretreatment rendered the graphite hydrophilic by introducing polar groups such as -OH and -COOH to the edge carbon atoms of the graphite sheets. Interaction of $POCl_3$ with these polar groups may have resulted in a strongly bonded residue of phosphoryl groups, as shown schematically in Figure 10, which effectively blocked the normally reactive sites at the edges of the graphite basal planes.

GRAPHITE	HYDROXYLATED GRAPHITE	PHOSPHORYLATED GRAPHITE
OXIDATION TEMP: 770°C	740°C	900°C

Figure 10. Surface treatment of graphite: reaction of NaCl and $POCl_3$ with edge carbons of the graphite sheet (schematic) [71].

It is questionable whether additives, such as metal phosphates and borates, that smother the carbon surface with a glassy residue, can properly be regarded as oxidation inhibitors. Phosphates can, however, deactivate catalytic impurities in the carbon by converting them to stable phosphates which cannot participate in the oxidation-reduction cycles discussed in Section III.I [71,73].

H. Topographical Effects Induced by Catalysts

As noted above, localized pitting and channeling of the carbon substrate are commonly observed to accompany the catalyzed oxidation of carbon. Mobility of catalyst particles and the simultaneous channeling of the basal plane of graphite crystals are unique and somewhat mysterious features of graphite gasification reactions in the presence of active catalysts, and the phenomena merit further discussion. The effect was first studied by Hennig in 1962 [45] and has more recently been the subject of a number of detailed investigations [72,74] involving hot-stage microscopy [5,6] and CAEM [16,55]. In general, movement of the catalyst particles on the basal-plane surface of graphite begins at temperatures slightly below or coincident with the onset of the oxidation reaction, and most authors have assumed that the motion is associated with the catalytic process itself. However, mobility of the particles on graphite has been observed in an atmosphere of CO when, presumably, no gasification reaction was occurring [15]. Also in a recent CAEM study, Baker and Skiba [19] observed mobility of Ag particles on graphite basal-plane surfaces when heated in dry helium or hydrogen. Although no etch pits, channels, or other signs of gasification were detected in these environments, motion of 4-8 nm Ag particles was observed to begin at temperatures in the 320-350°C range. As the Tamman temperature for silver is 368°C, it is likely that the initiation of particle mobility was associated in this case with the onset of mobility of lattice atoms and was not directly related to a catalyzed gasification process.

The details of the particulate motion during catalytic gasification of graphite are often complex and variable from one catalyst to
another. It is usually found that the smallest particles exhibit the
most vigorous mobility, with particles above about 10 μm in size being
generally immobile and catalytically inactive [5]. In some cases,
particles may collide and coalesce and thereby lose some or all of
their activity and mobility [74]. The rapidity of motion shows a
strong temperature dependence, which is generally reversible on rais-
ing or lowering the temperature [6]. However, in some cases the
mobility of the particles, and hence the catalytic activity, diminishes slowly with time as the particles become converted to an inactive chemical state [72]. This loss in activity is more common in
the case of the catalyzed graphite-CO_2 and graphite-H_2O reactions
(Sections IV and V). In many cases the catalyst particles resemble
liquid droplets [46], even at temperatures hundreds of degrees below
the melting point (e.g., Pd, Co), whereas low-melting oxides will
form a well-defined liquid phase (e.g., V/V_2O_5, Mo/MoO_3) [17]. In
still other cases (e.g., Cu), the particles retain their initial
solid, irregular outlines during their motion on the graphite surface
[5]. Movement of the particles and the resulting channels often
follow well-defined crystallographic directions in the graphite basal
plane and, sometimes, channeling may occur mainly along steps and
edges in the surface. On the other hand, channeling is completely
irregular in some instances [5] and the moving particles may undergo
abrupt and frequent changes in direction. Exfoliation of the graphite
layers occasionally occurs [16], an effect which may be due to intercalation of catalyst species between the graphite layer planes, resulting in considerable swelling in the c-axis direction. It has
often been found that channeling is initiated only when the catalyst
particle comes into contact with a step on the graphite surface [41,
45]. Generally channels tend to increase in width with distance from
the active particle at the growing tip, as a result either of uncatalyzed oxidation at the edges of the channel or because of catalyzed
oxidation induced by a film of active material left on the channel

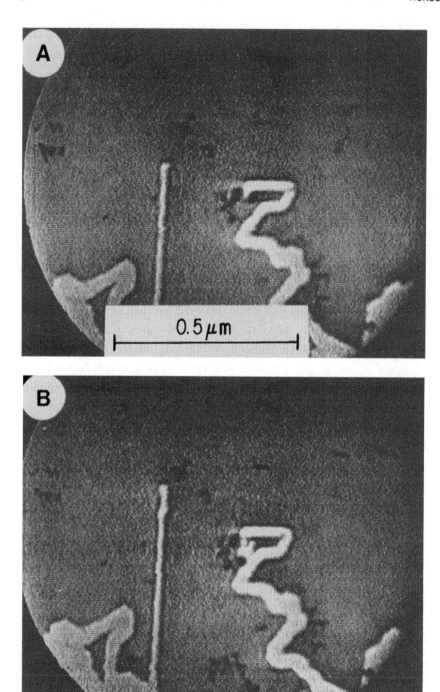

Figure 11. Sequential micrographs showing the propagation of channels formed by Ir particles on a graphite crystal surface on heating in 5 torr O_2 at 1000°C (Courtesy R. T. K. Baker).

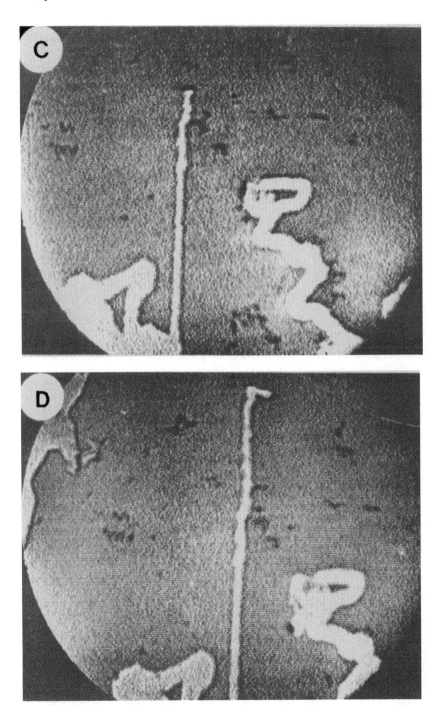

Figure 11   (Continued)

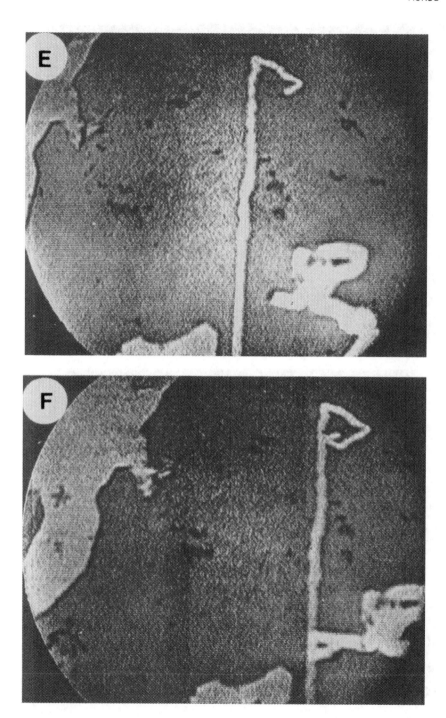

Figure 11   (Continued)

sides [16]. A sequence of microphotographs obtained by CAEM illustrating the propagation of channels formed by iridium particles is
shown in Figure 11. Typical channels formed by Co/oxide and V/V_2O_5
particles during the oxidation of graphite crystals are illustrated
in Figures 12 and 13.

On the basis of observations in the optical microscope, the
present author [6] classified the topographical effects produced on
flake graphite by metal or metal oxide particles into three general
categories, as follows:

1. Catalysts which caused erosion of the graphite surface in
   all directions, e.g., Pb
2. Catalysts which formed channels
   a. in preferred directions, e.g., V
   b. in random directions, e.g., Cu, Cd
3. Catalysts which formed etch pits
   a. hexagonal, preferentially in "perpendicular" orienta-
      tion, e.g., Fe
   b. irregular, in vicinity of particles, e.g., Mn, Ag

Some catalyst particles, however, produced channeling and etch pits
simultaneously in different regions of the graphite surface.

More recently, detailed studies of catalytic channeling have
been made using the CAEM technique for Ag [43], Mo [16], Pb [55],
Zn [53], and V [17]. A complex picture emerges from these studies,
and the behavior of catalyst particles has been shown to depend on
many factors such as particle size, oxygen partial pressure, and
temperature and heating rate in the microscope. Thus, with slow
heating rates in 4 torr (0.5 kPa) O_2 [43], Ag particles produced
mainly pits and channels that did not originate from edges on the
graphite surface, whereas at high heating rates only channels were
produced which started at surface edges. On the other hand, in the
hot-stage optical microscope [6], using 1 atm of O_2 (101 kPa), only
localized pitting was produced by Ag particles. Ag evaporated into
graphite flakes in the CAEM [43] and heated at rates slower than
$10°C \ sec^{-1}$ catalyzed the oxidation of the graphite at temperatures
above 365°C. A change in particle shape from spherical to hexagonal
was first observed, followed by a contraction of the particles to
reveal shallow hexagonal pits in the underlying graphite surface.

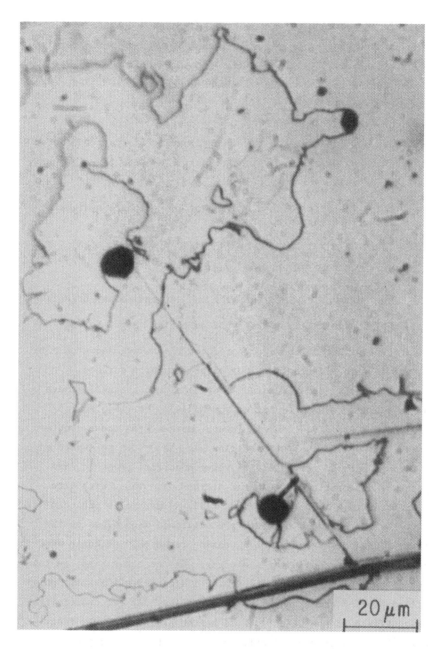

Figure 12. Catalytic channeling by Co/Co oxide particles on graphite during oxidation in dry O_2 at 750°C [6].

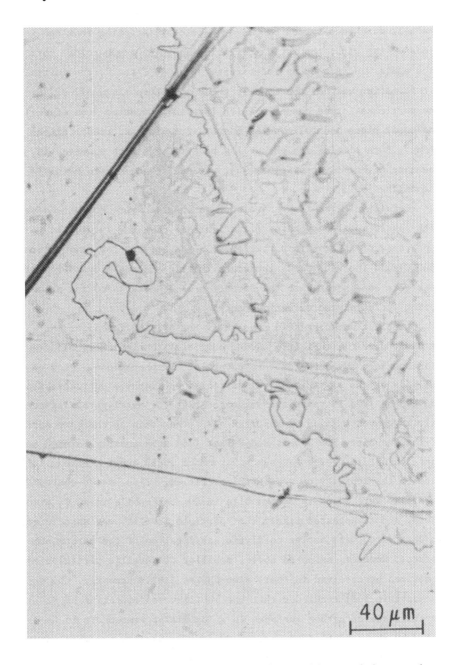

Figure 13. Catalytic channeling by vanadium oxide particles on the graphite basal plane during oxidation in dry O_2 at 600°C [6].

Channels were then formed by the Ag particles, starting from the
sides of the pits, the active particles assuming a spheroidal form
once again. With heating rates faster than $40°C \ sec^{-1}$, the initial
pit formation was not observed in the temperature range 477-927°C;
the particles formed channels which invariably started from edges in
the basal plane and followed in mostly irregular and random directions. Both pitting and channeling behavior was also observed in
the case of Mo/MoO_3 particles [16], whereas V/V_2O_5 caused mainly
channeling [6,17].

Catalytic channeling has also been observed with graphite doped
with alkali-metal carbonates [21]. In this case the liquidlike
appearance of the mobile particles at comparatively low temperatures
(600°C) and the results of thermogravimetric and thermodynamic analyses were consistent with the formation of a molten peroxide phase
during the catalyzed oxidation reaction.

Little new can be said concerning the driving force for the
catalyst particle mobility, and no completely satisfactory explanation has been advanced for this aspect of the phenomenon. From an
analysis of the migration of Ag particles on graphite surfaces, Sears
and Hudson [75] 15 years ago suggested that the metal particles actually move over a layer of adsorbed gas rather than on the bare surface
of the graphite, but Thomas and Walker [76] subsequently pointed out
that the nature of the gaseous environment could exert a strong influence on the mobility of some metal particles. Thus, oxide-coated Co
particles were found to be immobile in CO_2 but mobile in an O_2 atmosphere, and it appeared likely that particle mobility was associated,
in this case, with active catalytic gasification of the carbon substrate. However, as noted above, mobility of metallic particles may
sometimes be observed in inert atmospheres [19]. Recently, Chu and
Ruckenstein [48] concluded that the mobility and sintering of Pt
crystallites on carbon surfaces in an oxidizing atmosphere is facili-
tated by the evolution of gaseous CO and CO_2 beneath the metal particles. Although, in a general way, it appears likely that the evolution of gaseous oxides of carbon at the particle-graphite interface
during the oxidation reaction could help to explain the vigorous and

often irregular Brownianlike motion of the particles, the channeling behavior in preferred crystallographic directions is difficult to rationalize on this basis. Hennig [45] interpreted channeling behavior in terms of the formation of chemical bonds between the catalyst particle and reactive carbon atoms at the tip of the channel, but the vigorous mobility and the frequent changes in direction often observed with channeling particles are not readily explained on the basis of carbon-metal bond formation. Also, particles of alkali carbonates, which show very similar behavior [21], are unlikely to exhibit strong bonding to the carbon substrate. The generally liquidlike appearance of the moving catalyst particles is also something of a mystery. The heat liberated in the immediate vicinity of the particle by the strongly exothermic oxidation reaction could conceivably raise the particle temperature to a value above the Tamman point $(0.52T_m K)$ where liquidlike behavior would be expected to occur. However, in view of the high thermal conductivity of graphite in the basal plane it is surprising that particles of, for example, platinum (mp 1769°C) should behave as liquid droplets at catalytic oxidation temperatures as low as 500°C [46]. If localized heating at the reaction site is the cause of this phenomenon, then it should not be observed with the endothermic $C-CO_2$ and $C-H_2O$ reactions. However, catalytic channeling by iron particles has been observed on graphite surfaces during gasification in water vapor [77]. It has, nevertheless, been reported that particles of alkali-metal carbonates dispersed on graphite flakes show no mobility or channeling when heated in CO_2 to temperatures of up to 1000°C [21].

Although the occurrence of channeling in preferred crystallographic directions certainly suggests anisotropy of the catalyzed oxidation reaction in different directions within the graphite basal plane, it has not proved possible to quantify this effect or to interpret the observed differences in behavior between individual catalysts. The rule stated by Thomas [72] that the directions of highest reactivity are parallel to the faces of the growing etch pits can be used to interpret the directions of preferred reaction for catalysts that promote hexagonal pit formation. Thus, Mo/MoO_3

droplets which promote the formation of "parallel" pits [16] must
accelerate oxidation in the $< 10\bar{1}0 >$ directions, whereas Fe/oxide
particles which promote the growth of "perpendicular" pits [6] bring
about preferential oxidation of the graphite in the $< 11\bar{2}0 >$ directions. The reasons for this difference in behavior is, however,
unknown at the present time.

Although all the observed topographical features of catalytic
gasification are very hard to rationalize, the intensely localized
action of the catalyst particles is evident from these studies. It
follows that the proximity, mode of contact, and physical form of the
catalyst particles and carbon substrate are of paramount importance
in determining the magnitude and topographical details of the catalytic process. In addition, enhanced local gasification rates are
also observed when the impurities are finely dispersed in the bulk of
the carbon rather than on its surface. Adair et al. [9] doped glassy
carbons prepared from polyfurfuryl alcohol with fine dispersions of a
number of active metals (e.g., Ag, Cu, Ni, Fe) and then examined the
specimens by scanning electron microscopy after partial gasification
in O_2, CO_2, and N_2O. In general, although gasification of the pure
glassy carbon was very uniform, in the presence of added impurities
reaction was restricted to well-defined localized areas. Although
it continues to seem unlikely that the metallic additions were atom-
ically dispersed throughout the carbon matrix, the areas of preferred
attack very definitely corresponded to the regions in the immediate
vicinity of the metal inclusions. Diamonds containing transition
metal impurities show similar behavior [56]. Figure 14A shows a
crystal of diamond containing nickel impurity which had subsequently
been etched in oxygen at 600°C. It is obvious that oxidation was not
uniform; localized gasification in the vicinity of Ni inclusions was
responsible for the honeycomb structure shown. This behavior is
quite different from that of higher-purity diamonds on oxidation.
In this case, as shown in Figure 14B, oxidation was accompanied by
the formation of surface etch pits (trigons) at emerging dislocation
cores on the octahedral crystal faces. This behavior is analogous

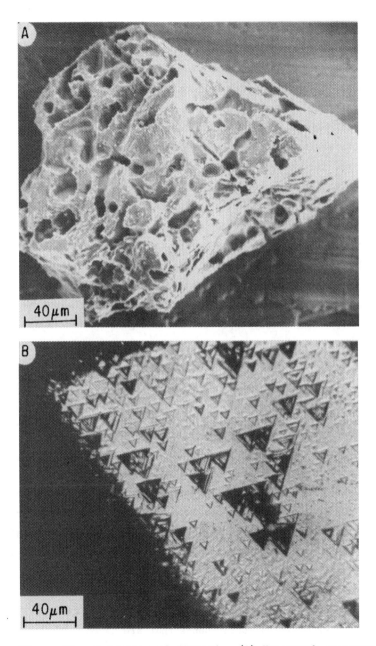

Figure 14.   Oxidation of diamond.   (A) Honeycomb structure produced
in diamond containing Ni inclusions following oxidation in O_2 at
700°C.   (B) Etch pits (trigons) formed on surface of higher-purity
diamond during oxidation in O_2 at 1000°C [56].

to the formation of hexagonal etch pits on graphite basal-plane surfaces during oxidation in the absence of catalysts [72].

Mobility of catalyst particles may also play a role in determining the changes in physical properties that often accompany the gasification of impure carbons. The rapid increase in porosity found after small amounts of gasification of microcrystalline-graphite compacts may be influenced by the migration of impurity particles into the interior, leading to increased local gasification rates and the formation of pores at distances remote from the surface. A technological application can be found in the area of carbon brush-wear at elevated temperatures, for copper or copper oxide particles originating from the commutator or slip-ring are frequently found at considerable distances from the sliding surface of the brush following a period of oxidative wear [50]. Catalytic oxidation of the brush can lead to the development of internal porosity in the neighborhood of the migrating catalyst particles and a general weakening of the brush structure.

I. Mechanism of the Catalyzed C-O_2 Reaction

Two general mechanisms have been proposed in the past to account for the diversified catalytic effects of metals, oxides, and salts in carbon oxidation reactions. These theories can be broadly classified as electron-transfer and oxygen-transfer mechanisms, and as they have been described in detail in previous reviews [3,4] only the main features of each will be summarized here. However, recent investigations that have shed light on the probable mechanism of the catalytic action of individual additives will be described in some detail.

The electron-transfer theory, proposed about 30 years ago [78], is based on the observation that many of the known catalytically active additives have unfilled energy bands that are capable of accepting electrons from the carbon matrix or, alternately, possess labile electrons that can be donated to the carbon. This electron transfer is assumed to result in a redistribution of π-electrons, a weakening of the carbon-carbon bonds at edge sites of the graphite sheets, and an increase in carbon-oxygen bond strength during the catalyzed oxidation reaction. This effect could be significant in determining the rate of desorption of CO, a process which has often been assumed to be the rate-determining step in the oxidation reaction [79].

According to Heuchamps [80], the chemisorption of oxygen is controlled by the charge on an electrical double layer at the carbon surface, and impurities that influence the kinetics of carbon oxidation can be divided into three classes: (1) electron donors, such as alkali metals, which form positive ions at the carbon surface, decrease the potential energy barrier to oxygen chemisorption, and cause increased rates of carbon oxidation; (2) electron acceptors, such as halogens, which form negative ions at the surface, increase the potential energy barrier, and act as oxidation inhibitors; and (3) transition metals, which accept electrons into their unfilled d-bands and thereby produce active sites on the carbon surface and increase the rate of oxidation. However, attempts to identify simple correlations between catalytic activity and properties such as ionization energy or lattice energy have not been very successful. An earlier correlation between activation energy for the catalyzed oxidation reaction and the lattice energy of the monoxides of the transition metals [81] has recently been disputed [82].

A variant of the electron-transfer mechanism is that proposed by Franke and Meraikib [23] to account for the catalytic effects of the alkali-metal carbonates in both the $C-O_2$ and $C-CO_2$ reactions. At temperatures in the vicinity of the melting point, group IA carbonates decompose slightly to yield small amounts of the alkali-metal monoxides. For example, the decomposition of Na_2CO_3,

$$Na_2CO_3 = Na_2O + CO_2$$

occurs to the extent of about 1% at the melting point of Na_2CO_3 i.e., 841°C [83]. Some dissociation of the oxide,

$$Na_2O = 2Na + \frac{1}{2}O_2$$

is also possible at elevated temperatures. Franke and Meraikib suggest that during the catalytic oxidation of graphite in the presence of Na_2CO_3, Na vapor intercalates into the graphite lattice and acts as an electron donor (alternatively halogens would act as electron acceptors). Alkali-metal-intercalated graphites show greater electron mobilities and higher electronic conductivities than pure graphite [84]. Such doping of the graphite lattice with electron donors would be expected to influence the chemisorption and desorption of oxygenated species and thereby promote the electron-consuming steps involved in the gasification reaction, e.g.,

$$e^- + CO_2(g) = CO_2^- \text{ (ads)}$$

or

$$e^- + CO_2(g) = CO(g) + O^- \text{ (ads)}$$

Similarly the chemisorption of O_2 or H_2O would be expected to be enhanced by the presence of electron donors such as alkali metals. However, although Na intercalation in graphitized carbons is measurable at 600°C [85], a typical catalytic gasification temperature, the equilibrium constant for the reaction,

$$Na_2O(\ell) = 2Na(g) + \frac{1}{2}O_2(g)$$

is only about 2×10^{-20} at 600°C, and Na_2O is only measurably reduced by carbon at temperatures in excess of 800°C [21]. It appears unlikely, therefore, that intercalation of graphite by Na would be significant during catalytic oxidation at 600°C or below. It has, however, been demonstrated that alkali-metal vapor is formed during the catalytic gasification of graphite by CO_2 at temperatures around 1000°C [21], and this mechanism may play a major role in this case (Section IV.C), although it is not clear that the Na-graphite intercalation compound will form at such high temperatures.

The oxygen-transfer mechanism regards active catalysts for carbon oxidation as oxygen carriers that undergo oxidation-reduction cycles on the carbon surface during the course of the reaction. In other words, an intermediate compound, such as a metal oxide, is formed in the presence of the oxidizing gaseous environment; the oxide then is reduced on contact with the carbon substrate and the cycle is repeated as the mobile particle migrates on the carbon surface. Historically, this is the oldest explanation of the catalytic effect, having been originally proposed in general form by Kröger and co-workers in 1931 [86]. Although it suffered an eclipse during the period when all heterogeneous catalytic reactions were being explained by electron-transfer mechanisms, the idea of a specific oxidation-reduction sequence as an explanation of the behavior of individual catalysts has returned to favor [6,21,26,27]. On the basis of their observations with a wide range of metal oxides, Amariglio and Duval [36] concluded that only those metals capable of existing in two oxidation states were active catalysts for carbon oxidation.

A comparison of the general level of catalytic activity of various metals and oxides for carbon oxidation with the free energies of formation of the oxides [6] indicates that oxides in the broad class of those capable of being reduced by carbon at temperatures in the range of 500 to 700°C (either to the metallic state or to a lower oxide) are effective catalysts; whereas more stable oxides such as those of Si, Al, W, and Ta are inactive. Simple free-energy considerations, however, provide little guide to the relative catalytic activity exhibited by the different metals and oxides, and exceptions to this generalization include Ag and, possibly, ZnO. Particles of the latter probably function as oxygen dissociation sites, as both Ag [87] and ZnO [88] are known to be active catalysts for isotopic oxygen exchange and for other oxidation reactions involving the fission of the O-O bond in molecular O_2. Mobile oxygen atoms thus formed on the metal (or oxide) surface could rapidly migrate to the metal (or oxide) carbon interface causing gasification of the carbon substrate in the vicinity of the catalyst particle.

An indication that the catalyzed oxidation process is essentially chemical in nature is the observation that the activity of metallic catalysts is very dependent on the nature of the oxidizing gas. Thus the effect of transition metals such as Fe is generally less marked in air or O_2 than when the oxidation is carried out in CO_2 [42]. On the other hand, some metals such as Ag and Cu that are very active in O_2 have virtually no effect on the kinetics of oxidation of graphite in CO_2 [42], presumably because these metals are much less capable of dissociating the C=O bond than the weaker O-O bond. Amariglio and Duval [36] found that the activation energy for the catalyzed oxidation of carbon was essentially independent of the concentration of catalytic impurity present, a result which would not be expected on the basis of electron transfer. However, conclusions based on apparent activation energies are notoriously unreliable because of the widely different trends reported in various studies, as noted below.

Thermal analysis, especially thermogravimetric and differential thermal analysis, has proved particularly useful in lending support to the oxygen-transfer mechanism by identifying the specific oxidation-reduction cycles involved with particular catalysts. Thus by carrying out separate TG-DTA measurements in both inert gas and in oxygen with mixtures containing equal-weight fractions of graphite and a catalytically active salt or oxide, the temperature at which the metal oxide begins to be reduced by the carbon has been shown in many cases to coincide with the temperature for the onset of the catalytic oxidation reaction [6]. For example, the oxidation of graphite is strongly catalyzed by small amounts of copper salts, as noted above. The addition of 0.01% of copper acetate to a typical graphite powder has been found to lower the ignition temperature of the graphite in oxygen from 740°C to about 600°C, and with higher concentrations of Cu, little additional reduction in ignition temperature is observed (Fig. 15). Copper acetate decomposes on heating in O_2 to yield CuO at a temperature of about 300°C. On heating, a mixture of CuO and graphite powder in nitrogen in the thermobalance at linearly increasing temperature reduction of the oxide to the metal began at a temperature of 600°C [5], as shown in Figure 16. It appears likely, therefore, that when graphite

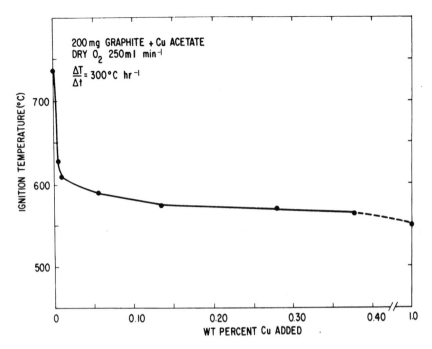

Figure 15. Effect of Cu concentration on the ignition temperature of graphite powder in O_2 [5].

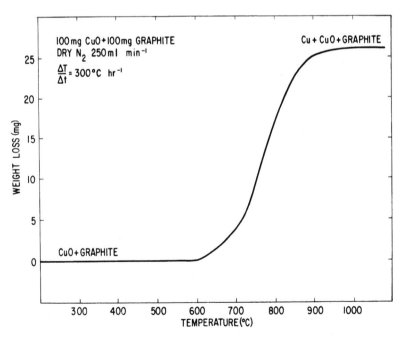

Figure 16. Thermogram for a 1:1 graphite-CuO mixture on heating in dry N_2 [5].

containing added copper acetate is heated in oxygen or air the decomposition of the copper salt at low temperatures is followed by
gasification of the carbon at 600°C and above, as a result of the
cyclic process

$$CuO + C = Cu + CO$$

$$Cu + \frac{1}{2}O_2 = CuO$$

This sequence of elementary steps has been confirmed by magnetic-
susceptibility measurements of the carbon-Cu system during catalytic
oxidation [7].

Similar considerations apply to many other catalyst systems [6].
For example, the oxides As_2O_3, Sb_2O_3, and Bi_2O_3, which are catalysts
for C oxidation, are all reducible to the metals on heating with
carbon at 500-700°C. In some cases the metal oxide is reduced by
the carbon substrate to a lower oxide, rather than to the metallic
state. Thus a 1:1 mixture of V_2O_5 and graphite, heated at a linearly
increasing temperature in flowing nitrogen in the thermobalance,
showed a sudden weight loss at 675°C (Figure 5) as a result of the
reduction of V_2O_5 to the lower oxide V_6O_{13} [6], and it is likely
that the cyclic catalytic oxidation process in this case involves
the sequential steps

$$3V_2O_5 + 2C = V_6O_{13} + 2CO$$

$$V_6O_{13} + O_2 = 3V_2O_5$$

By selective-area electron diffraction [17] it has also been shown
that an active catalyst particle that was originally V_2O_5 was converted to V_6O_{13} during catalytic oxidation on a single-crystal
graphite surface. On the other hand, evidence for the conversion
of V_2O_5 to V_2O_4 during the catalyst oxidation of carbon has been
obtained by electron-spin resonance [25]. The mechanism of catalysis
by the platinum-group metals is less certain. However, the identification of a platinum-oxide phase during the Pt-catalyzed oxidation of

carbon by O_2 [48] suggests that an oxidation-reduction cycle may be operative in this case also.

A recent study of the catalytic behavior of alkali-metal carbonates and oxides in graphite oxidation [21] has lent further support to the oxygen-transfer mechanism for this group of active materials. Although, as noted above, thermal dissociation of the pure group IA carbonates to the corresponding monoxides occurs to only a very small extent at temperatures in the vicinity of 500°C and only to the extent of about 1% at equilibrium in the neighborhood of the carbonate melting point [83], in the presence of carbon and oxygen the dissociation of the carbonate may be enhanced as a result of the sequence of reactions,

$$M_2CO_3 \rightarrow M_2O + CO_2$$
$$2M_2O + O_2 \rightarrow 2M_2O_2$$
$$2M_2O_2 + C \rightarrow 2M_2O + CO_2$$

as monoxide is converted by reaction with O_2 to peroxide, which subsequently reacts with the carbon substrate. The overall result will be an increase in the rate of carbon gasification and increased concentrations of oxides in the salt phase. Table 4 shows the amounts of alkali-metal oxides (monoxides + higher oxides) formed on heating

Table 4. Formation of oxides from alkali carbonates on heating
in presence of O_2 and graphite [21]

Carbonate M_2CO_3	M_2O conc. in original M_2CO_3 (wt %)	M_2O conc. after heating M_2CO_3 in O_2 for 4 hr at 900°C (wt %)	M_2O conc. after heating M_2CO_3 + graphite (1:1) in O_2 for 4 hr at 900°C (wt %)
Li	<0.2	2.1	4.6
Na	<0.2	0.2	1.1
K	<0.2	0.6	7.6
Rb	<0.2	1.6	6.6
Cs	<0.2	3.2	7.7

Source: From Ref. 21.

the carbonates both alone and mixed with graphite for 4 hr in oxygen
at 900°C. Although appreciable amounts of the oxides were formed by
decomposition of the carbonates at this temperature, the addition of
graphite produced substantial increases in the concentration of oxides
formed. On the other hand, alkali-metal sulfates and chlorides, which
are less active carbon oxidation catalysts than the carbonates, formed
only small concentrations of oxides under the same conditions.

As mentioned above, on heating Na_2O in oxygen at increasing temperatures a rapid reaction begins at 250°C with the formation of Na_2O_2
(Figure 4), which melts at 460°C and reacts vigorously with graphite
at this temperature. The observed catalytic channeling by liquid
droplets during the Na_2CO_3-catalyzed oxidation of graphite crystals
at 600°C may therefore result from the formation of Na_2O_2 on the
graphite surface during the oxidation reaction [21]. The catalytic
effect of the alkali-metal oxides probably involves an oxidation-
reduction cycle of the type,

$$M_2O + \frac{n}{2}O_2 = M_2O_{1+n}$$
$$M_2O_{1+n} + nC = M_2O + nCO$$

where a peroxide (e.g., Na_2O_2) or higher oxide (e.g., Rb_2O_3, CsO_2)
acts as a reactive intermediate species. $\Delta G°$ values for both these
reactions are negative for most M_2O and temperature conditions of
relevance, and both reactions are thermodynamically favored over a
wide range of P_{O_2} and P_{CO}. The energetics of peroxide formation are
actually more favorable with the K than with the Na species at temperatures in the 500-1000°C range so that K_2O would be expected to
be catalytically active over a wider range of P_{O_2} and P_{CO} conditions
than Na_2O, as found experimentally.

At higher temperatures or in the presence of a molten carbonate
phase, it is possible to represent the oxygen-transfer mechanism in
electrochemical terms [89], but a flow of electrons through the carbon
substrate must be assumed. Thus carbon, at an anodic site on the surface, is oxidized by the carbonate electrolyte,

$$C + 2CO_3^{2-} = 3CO_2 + 4e^-$$

and at a separate cathodic site, regeneration of carbonate occurs,

$$O_2 + 2CO_2 + 4e^- = 2CO_3^{2-}$$

giving the overall cell reaction,

$$C + O_2 = CO_2$$

In this mechanism the alkali-metal cation does not play a role and the catalytic activity differences between the individual carbonates are not explained. This electrochemical process has, however, been utilized in a molten-carbonate fuel cell to produce electricity from coal [89].

Although there have been a few reported instances where the presence of a catalyst increased the apparent activation energy for the $C-O_2$ reaction [81], many [3,54] have observed a decrease in activation energy upon the addition of a catalyst to a pure carbon. However, attempts to correlate activation energies with atomic number of the catalyzing metal have not been notably successful [82]. It is extremely difficult to prove that an observed decrease in activation energy is not due to partial control of the kinetics of the reaction by diffusion processes taking place in the developing pore system of the gasifying carbon. Cases have, in fact, been reported [90] where the activation energies for the catalyzed and uncatalyzed gasification reactions were the same, in spite of large differences in the overall rates. This observation implies that the effect of catalysts is to increase the preexponential factor of the rate equation, which would be the expected result of an increase in the density of reaction sites on the carbon surface. On the other hand, a real reduction in activation energy, as distinct from a decrease due to mass transport limitations, would imply that the catalyst effectively increases the rate of the limiting step in the reaction sequence. Either effect is thus

possible and can be rationalized. However, it is often found that
the values of the preexponential factor and the apparent activation
energy change together in the same sense, according to a compensation
effect [27,28,54]. Thus, as with many other catalyzed reactions [87],
the kinetic parameters of the catalyzed oxidation of a series of
carbons may fit a relation of the form,

$$\log A = \log k_o + \frac{E}{RT_s}$$

where E is the apparent activation energy, A the corresponding preexponential factor, and T_s the temperature (isokinetic) at which all
the reactions of the series proceed at the same rate. If a compensation effect is operating, and the activation energy of the uncatalyzed
reaction is greater than that of the catalyzed reaction, an active
catalyst at, for example, 500°C must have an even greater isokinetic
temperature. The isokinetic temperature for Zn on graphite has been
estimated to be 1097°C [53] so that Zn will only show significant
catalytic activity at temperatures below this point.

Feates et al. [91] have developed an alternative model for the
compensation effect for carbon oxidation reactions in which catalyzed
and uncatalyzed reactions are assumed to take place simultaneously and
independently on different parts of the carbon surface. If these two
reactions have kinetic parameters A_c, E_c and A_u, E_u, respectively, the
total gasification rate will be given by

$$k = sA_c \exp\left(\frac{-E_c}{RT}\right) + (1 - s)A_u \exp\left(\frac{-E_u}{RT}\right)$$

where s is the fraction of the active carbon surface on which the
catalyzed reaction is occurring or the fraction of the reaction rate
due to the catalyzed reaction. If the reactions exhibit a compensation effect, at the isokinetic temperature T_s

$$k_o = A_c \exp\left(\frac{-E_c}{RT_s}\right) = A_u \exp\left(\frac{-E_u}{RT_s}\right)$$

$$\frac{k_c}{k_u} = \frac{A_c}{A_u} \exp\left(\frac{E_u - E_c}{RT_s}\right) = 1$$

hence

$$k = k_o [sZ_c + (1 - s)Z_u]$$

where

$$Z_i = \exp\left[\frac{E_i(1/T_s - 1/T)}{R}\right]$$

when $s \gg 1$,

$$\log k \approx \log k_o + \frac{E_u}{R}\left(\frac{1}{T_s} - \frac{1}{T}\right)$$

and when $s = 1$,

$$\log k = \log k_o + \frac{E_c}{R}\left(\frac{1}{T_s} - \frac{1}{T}\right)$$

Arrhenius plots of log k vs. $1/T$ are shown schematically in Figure 17 (a) and (b) for the cases (a) $E_u > E_c$ and (b) $E_u < E_c$.   It

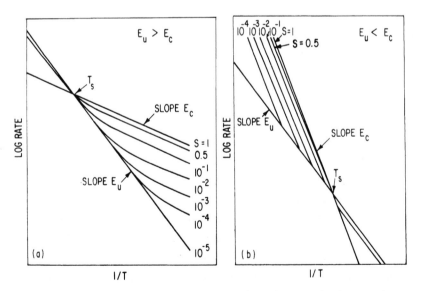

Figure 17.   Calculated values of log (rate) vs. $1/T$ for various values of s for (a) $E_u > E_c$, and (b) $E_u < E_c$ (according to model of Feates, Harris, and Reuben [91].

is evident that for most practical cases when s is small, the observed
activation energy will correspond to that for the uncatalyzed reaction
E_u at temperatures more than a few degrees above T_s when $E_u > E_c$ or at
temperatures less than a few degrees below T_s when $E_u < E_c$. In the
latter case when $T > T_s$, increasing burnoff will tend to concentrate
the catalyst on the active carbon surface, the value of s will increase, and the observed activation energy will approach E_c. The same
increase in apparent activation energy should be observed on increasing the initial concentration of catalyst. Similar effects would be
expected when $E_u > E_c$ at oxidation temperatures less than T_s. Examples
are cited from the literature of both types of behavior to support this
model. However, as the kinetics of catalyzed gasification reactions
of carbon do not always exhibit compensation effects [90], it is not
clear that this model has general validity.

IV. CATALYZED C-CO$_2$ REACTION

A. Introduction

The reaction of carbon with CO_2 (the Boudouard reaction) [26],

$$C(graphite) + CO_2(g) = 2CO(g) \quad \Delta H°_{298.15} = +41.21 \text{ kcal mol}^{-1}$$
$$\Delta G°_{1200 \text{ K}} = -9.42 \text{ kcal mol}^{-1}$$

is endothermic and, for a given carbon in the absence of a catalyst,
takes place several orders of magnitude slower than the $C-O_2$ reaction
at the same temperature [2]. In general, catalysts are needed to
produce substantial gasification rates for the $C-CO_2$ reaction at
temperatures below 900°C. The reverse reaction, involving the dissociation of CO, takes place at temperatures as low as 500°C on many
catalytically active metal surfaces [92].

Although the Boudouard reaction always proceeds to some extent
during the steam gasification of coal, the effects of catalysts on
the kinetics have generally received much less attention than the

$C-O_2$ reaction. These oxidation reactions are often discussed together, and similar catalysts are often effective in both cases. However, the catalyzed Boudouard reaction exhibits some unique features and the detailed mechanisms of the catalytic effect are probably distinct from those involved in the $C-O_2$ system.

B. Catalysts

In general the most active catalysts for the Boudouard reaction are the salts of the alkali and alkaline earth metals and the metals of the iron and platinum group.

The carbonates and oxides of the group IA metals have been known for a long time to be effective in accelerating gasification rates of carbon in CO_2, the salts of lithium being generally the most active. Figure 18 summarizes weight-loss measurements for graphite-alkali carbonate mixtures (1:1 by weight) on heating in flowing CO_2 in a thermobalance at a linearly increasing temperature of 10°C min^{-1} [21]. In this series of experiments, gasification began at temperatures slightly in excess of 700°C, and the reaction was accompanied by some vaporization of alkali metal into the gas stream. In separate experiments with graphite single crystals coated with dispersed alkali-metal carbonates and mounted in a hot-stage microscope, the author observed no catalytic channeling or particle mobility at temperatures up to 1000°C in a CO_2 atmosphere. This is in contrast to the behavior of the same alkali carbonates during the gasification of graphite in molecular O_2, in which case catalytic channeling by mobile liquid droplets was readily observed at temperatures as low as 600°C [21].

The kinetics of the Boudouard reaction with a carbon black doped with 1.5 and 3.7 wt % Li_2O and Li_2CO_3 have recently been measured in the temperature range 839-1050°C [93]. As alkali-metal monoxides react with CO_2 at low temperatures, it is likely that Li_2CO_3 was the actual species present at the Boudouard reaction temperature in both cases. Indeed, the kinetics were similar whether Li_2O or Li_2CO_3 were added initially. After correction of the experimental rate data for pore

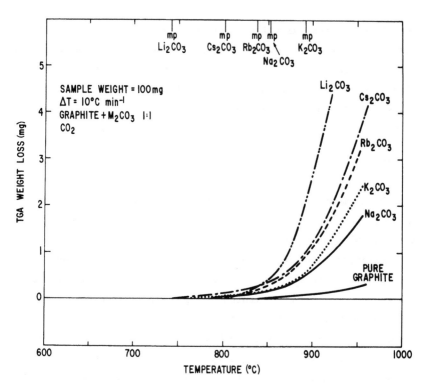

Figure 18. Effect of alkali-metal carbonates on the gasification of
graphite powder in CO_2. Weight changes vs. temperature. Heating
rate = 10°C min^{-1}. Gas flow rate = 400 ml min^{-1} [21].

diffusion and film transfer effects, the intrinsic rate constant
(K_v), in mol cm^{-3} atm^{-1} sec^{-1}, of the catalyzed reaction for both
concentrations of additive was found to follow the relation,

$$\log K_v = 5.471 \ (\pm 0.353) - \frac{48.040}{2.303RT}$$

Lithium oxide has also been shown to be effective in accelerating the reduction of hematite by carbon [94], presumably because of
the activity of Li_2CO_3 as a catalyst in the Boudouard reaction in the
sequence of reactions,

$$Fe_xO_y + CO = Fe_xO_{y-1} + CO_2$$
$$C + CO_2 = 2CO$$

The gasification of metallurgical coke by CO_2 at 900°C has recently been shown to be strongly catalyzed by potassium salts [95], the catalytic activity decreasing in the order K_2CO_3 > KCl > K_2SO_4 > K_3PO_4.

There is also evidence that oxides and carbonates of the alkaline earth metals are active catalysts for this reaction. Thus Hippo and Walker [96] have observed that Mg and Ca salts present in coal char enhance the reactivity toward CO_2 at 900°C. Also, a marked decrease in the reactivity of oil-shale char toward CO_2 takes place when the char is acid-leached to remove carbonate minerals such as calcite and dolomite [97]. In a recent study by the author [98] in which pure graphite powder was doped with alkaline earth carbonates and the kinetics of gasification in CO_2 measured between 700 and 1000°C, the order of catalytic activity was found to be $BaCO_3$ > $SrCO_3$ > $CaCO_3$ > $MgCO_3$. The addition of 5 wt % $BaCO_3$ increased the gasification rates of the graphite by more than 3 orders of magnitude at 900°C.

A detailed study of the behavior of iron and its oxides as catalysts in the Boudouard reaction has been described in a previous review [3]. In this definitive investigation utilizing magnetic-susceptibility measurements, Walker and co-workers have demonstrated that the catalytic activity of the metal declines sharply as the metallic phase is oxidized progressively to wüstite $Fe_{0.95}O$ and then to magnetite Fe_3O_4, which is completely inactive. However, an inactive, oxidized-iron catalyst could be reactivated by reduction by CO, by H_2, or even by carbon at high temperatures. Similar results were obtained by Turkdogan and Vinters [42] for iron, cobalt, and nickel catalysts added to electrode graphites. In this case the addition of less than 0.01% iron increased the gasification rate of the graphite in CO_2 by a factor of 2000 in the temperature range 700–1000°C, whereas 2% iron increased the rate by a factor of about a million at 800°C. Also in this latter study, silver and the oxides of Cu, Zn, and Cr, which are active catalysts for carbon gasification in O_2, were found to be inactive in CO_2. However, Adair et al. [9] found that Ag and Cu particles induced localized pitting of polymer-derived carbons on heating in CO_2 and hence presumably accelerated the overall gasification rate.

As with the other catalyzed gasification reactions, metallic particles induce an intensely localized effect during the gasification of carbon in CO_2, producing pitting and channeling in the carbon substrate [9,11,99]. The topographical changes occurring in polymer-derived carbons during the catalyzed Boudouard reaction have been intensively studied by Marsh and co-workers [9,11]. For example, on heating a polyfurfuryl alcohol glassy carbon containing finely dispersed Fe and Ni particles in CO_2 at 850°C, gasification was observed to take place preferentially in the immediate vicinity of the metallic

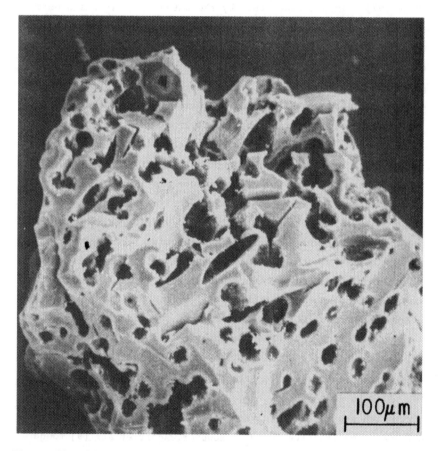

Figure 19. Pitting produced in polyfurfuryl alcohol carbon by Fe particles, after 72% burnoff in CO_2 at 790°C [9].

inclusions to produce a honeycomb structure of macropores in the
carbon matrix, as illustrated in Figures 19 and 20. Catalytic
channeling and mobility of the metallic particles was also observed.
Other authors [99] have also reported a large increase in porosity
when Ni-impregnated carbons are heated in CO_2 at, for example, 600°C.

To a greater extent than with the catalyzed gasification of
carbon by molecular O_2, the occurrence of the Boudouard reaction is
often accompanied by a gradual loss of activity of the catalyst particles, a process which appears to be associated with the conversion

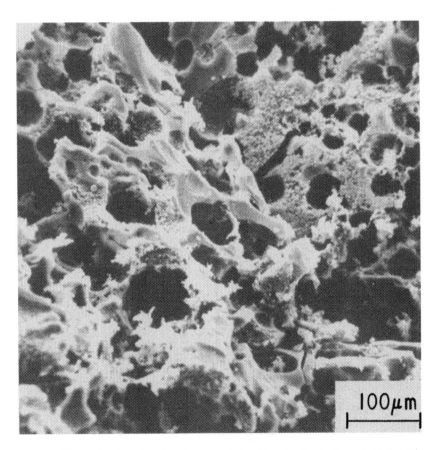

Figure 20. Pitting produced in polyfurfuryl alcohol carbon by Ni
particles, after 61% burnoff in CO_2 at 550°C [9].

of the initially active metallic particles to an inert oxide phase
[3]. Thus, in the study by Marsh and Rand [11], both Ni and Fe
particles lost their activity after an initial period of rapid gasification, as a result of the conversion of the metals to NiO and
Fe_2O_3, respectively. As in the study by Walker et al. [3], subsequent heating of the deactivated catalysts in hydrogen at 800°C for
3 hr reduced the oxides to the metallic state, and the catalytic
activity was restored.

Recently, Tashiro et al. [100] have shown that noble metals of
the platinum group are also active catalysts for the $C-CO_2$ reaction,
with the highest activity being exhibited by Rh. The activity of Pd
decreased rapidly with time at 800°C.

Tamai et al. [101] recently examined the reactivity toward
CO_2 of an activated carbon that had been impregnated with group
VIII metals to give 4.8% metal by weight. The reaction rates as a
function of temperature are shown in Figure 21 for the various metal
additives and also for the original carbon without added catalyst.
Ni, Ru, Rh, and Os showed a peak in reactivity at temperatures
in the range 500-800°C and increasing reaction rates at higher

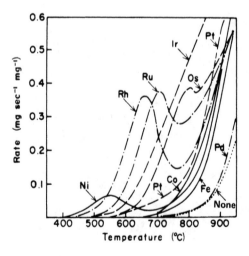

Figure 21. Catalytic gasification of active carbon in CO_2. Rates
vs. temperature for various catalysts [101].

temperatures. Ir and Pt exhibited a high catalytic activity but no
low-temperature peak was observed. Fe and Co were moderately active
catalysts at the higher temperatures, but Pd was inert. Similar
patterns of activity were observed with the same impregnated carbon
samples when gasified in steam or H_2, suggesting that the low-
temperature reactivity was due to the selective catalytic gasification of a reactive component in the carbon sample, although the
reason for the different behavior of the various metals is not clear.

Little information is available concerning inhibitors of the
$C-CO_2$ reaction, although it would be expected that strongly chemisorbed species such as $POCl_3$ and halogens, which inhibit the combustion of carbon in O_2, would be effective in this case also. Allardice
and Walker [60,61] have found that substitutional doping of natural
graphite crystals with boron, at the 0.2 and 1.0 at. % level, reduces
the subsequent rate of gasification in CO_2 but does not change the
activation energy for the overall reaction. These authors suggest
that in some undefined manner substitutional B retards the initial
chemisorption and oxygen-exchange step and hence retards the overall
gasification rate.

C. Mechanisms of the Catalyzed $C-CO_2$ Reaction

As with the catalyzed gasification of carbon by molecular oxygen,
several distinct mechanisms have been proposed to account for observed
catalytic effects in the Boudouard reaction. In an earlier review,
Walker et al. [3] have discussed the salient features and relative
merits of the electron-transfer mechanism (involving transfer between
the π-electrons of graphite and vacant orbitals of the metallic catalyst) and the oxygen-transfer mechanism (involving a cyclic oxidation-
reduction sequence of elementary steps). In the case of catalysis by
metallic iron, these authors have presented a strong case for a cyclic
oxidation-reduction process involving the initial dissociative chemisorption of CO_2 at the Fe surface to produce a labile chemisorbed
oxygen atom,

$$CO_2 + Fe = Fe - O_{ads} + CO$$

The chemisorbed oxygen atoms may then diffuse to the metal-carbon interface and there react with the carbon to yield gaseous CO,

$$Fe - O_{ads} + C = Fe + CO(g),$$

giving the overall reaction,

$$CO_2 + C = 2CO$$

and localized gasification of the carbon in the immediate vicinity of the metallic particles. There is separate evidence that the chemisorption of CO_2 on iron surfaces is dissociative [102] and that chemisorbed oxygen atoms on metal surfaces are mobile at elevated temperatures [103]. As the oxides of iron are unlikely to be effective in dissociating CO_2, the observed loss of catalytic activity for the Boudouard reaction as the iron is converted to wüstite or to magnetite as a result of incorporation of part of the chemisorbed oxygen into the metallic lattice, can be rationalized.

Grabke [104] has reported a marked change in the kinetics of the Boudouard reaction in the presence of iron at a gas composition corresponding to $P_{CO_2}/P_{CO} = 0.65$, which can be related to the Fe/FeO transformation. In this study the gasification reaction was considered as a sequence of the elementary steps,

$$CO_2(g) = CO(g) + O_{ads} \qquad \text{rate } v_1$$

and

$$O_{ads} + C = CO(g) \qquad \text{rate } v_2$$

By measuring both the rate of formation of [14]CO from [14]CO_2 and the overall gasification rate, the forward reaction rates v_1 and v_2 for these steps were estimated as functions of the ratio P_{CO_2}/P_{CO} in the gas phase. For an electrode graphite doped with Fe, the results shown in Figure 22 were obtained for the $C-CO_2$ reaction at 785°C. An abrupt change in kinetics as $P_{CO_2}/P_{CO} = 0.65$ was observed. At lower CO_2/CO ratios where metallic Fe is stable, the oxygen transfer rate v_1 was inversely proportional to the oxygen activity $a_o = P_{CO_2}/P_{CO}$ $\propto (P_{O_2})^{1/2}$, whereas v_2 was independent of oxygen activity, suggesting that the available sites on the Fe surface were saturated with chemisorbed oxygen. At higher CO_2 ratios, where FeO is stable, both

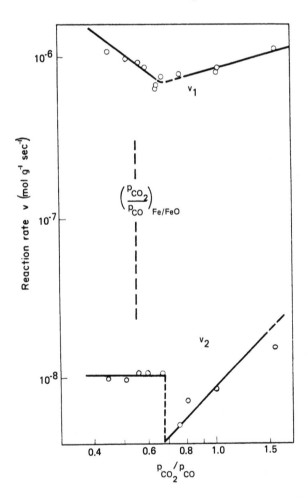

Figure 22. Plot of log (rate) vs. log (P_{CO_2}/P_{CO}) for an Fe-doped electrode graphite gasified in CO_2 at 785°C [104]. The equilibrium ratio P_{CO_2}/P_{CO} for the Fe → FeO transformation at 785°C is indicated.

v_1 and v_2 increased with increasing a_o, a result that was left unexplained.

The fact that vaporization of alkali metal has often been observed to accompany the $C-CO_2$ reaction when catalyzed by alkali-metal carbonates, suggests that an oxidation-reduction cycle involving the intermediate formation of free alkali metal might be occurring. A possible cycle might consist of the following sequential steps [21]:

$$M_2CO_3 + 2C = 2M + 3CO \tag{1}$$

$$2M + CO_2 = M_2O + CO \tag{2}$$

$$M_2O + CO_2 = M_2CO_3 \tag{3}$$

Although reaction (1) has a positive value of $\Delta G°$ at temperatures in the gasification range, significant pressures of alkali-metal vapor may exist in the system depending on the ambient pressure of CO, as for reaction (1),

$$K_1 = 3 \log P_{CO} + 2 \log P_M$$

Figure 23 shows the equilibrium values of P_{CO} and P_M corresponding to reaction (1) at 827°C (1100 K) for M = Li, Na, and K. At this temperature K, Rb, and Cs are in the gaseous states and Li and Na have appreciable vapor pressures. For an ambient pressure of CO of 0.1 atm (10.1 kPa), the vapor pressure of Na or K in the reaction environment could attain values of from 10^{-3} to 10^{-4} atm (101-10.1 Pa), which would be adequate to sustain an oxidation-reduction cycle of the type (1) → (2) → (3) → (1). Reactions (2) and (3) are likely to be rapid and reaction (1) is probably the rate-determining step of the sequence. The fact that Li_2CO_3 has the lowest melting point of the series of alkali-metal carbonates would help to promote the occurrence of reaction (1) in the case of Li, by increasing the salt-carbon interfacial contact area, even though the vapor pressure of this metal is lower than that of the other alkali metals. It has been suggested

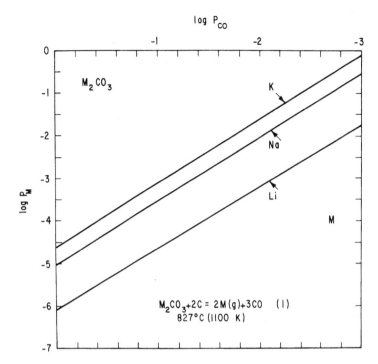

Figure 23. Equilibrium values of P_M and P_{CO}, corresponding to reaction (1) 827°C (1100 K) for M = Li, Na, and K [21].

recently [95] that alkali metal liberated by reaction (1) above can attack the carbon substrate causing extensive cracking, possibly as a result of a charge-transfer effect which weakens the carbon-carbon bonds.

Recent work [98] has suggested that an analogous mechanism applies to catalysis by the alkaline earth carbonates. As the group IIA oxides are more stable at high temperatures than are the monoxides of the group IA alkali metals, reaction of the carbonates with carbon proceeds only as far as the oxides in the former case. A two-step cycle of the type

$$BaCO_3 + C = BaO + 2CO$$
$$BaO + CO_2 = BaCO_3$$
$$\overline{\quad C + CO_2 = 2CO \quad}$$

is feasible on thermodynamic grounds. The individual reaction steps
have been found to occur readily in the 800-1000°C temperature range
with the first process being the rate-determining step in the overall
catalyzed gasification reaction. Because of the limited thermal
stability of Mg and Ca carbonates in this temperature range, the
catalytic activity of these salts is much less marked than that of
Sr and Ba carbonates. It may be noted parenthetically that the
modest catalytic effect of alkali-metal carbonates in the carbon-SO_2
reaction [29] may be due to the occurrence of an analogous oxidation-
reduction cycle, namely,

$$Na_2CO_3 + SO_2 = Na_2SO_3 + CO_2$$
$$2Na_2SO_3 + SO_2 = 2Na_2SO_4 + S(g)$$
$$Na_2SO_4 + 2C = Na_2S + 2CO_2$$
$$Na_2S + 2SO_2 = Na_2SO_4 + 2S(g)$$

the last two steps giving the overall reaction,

$$C + SO_2 = CO_2 + S(g)$$

Other roles have been suggested for alkali metal in the carbonate-
catalyzed Boudouard reaction. Franke and Meraikib [23] suggest that
Na intercalates into the graphite lattice and acts as an electron
donor. Doping of the graphite with an electron donor will promote
the chemisorption and dissociation steps,

$$e^- + CO_2(g) = CO_2^- (ads)$$

and

$$e^- + CO_2(g) = CO(g) + O^- (ads)$$

and thereby accelerate the overall gasification rate. However, in
view of the known thermal instability of Na-graphite intercalation
compounds [105], it seems unlikely that appreciable Na can enter

the graphite lattice at gasification temperatures in the neighborhood
of 1000°C.

An "electrochemical" mechanism has recently been proposed to
account for the catalytic effects of alkali-metal carbonates in the
Boudouard reaction. According to Jalan and Rao [93], alkali carbonate
above its melting point will form a thin layer of electrolyte over the
carbon surface. At anodic sites on the surface, carbonate ions react
with the carbon, evolving CO,

$$CO_3^{2-} + 2C = 3CO(g) + 2e^-$$

Alkali-metal cations M^+ then migrate through the melt and electrons
transfer through the carbon substrate to a neighboring cathodic site
where reduction of CO_2 takes place

$$2M^+ + CO_2(g) + 2e^- = M_2O(s) + CO(g)$$

and the resulting alkali-metal oxide is converted to carbonate by
reaction with ambient CO_2

$$M_2O(s) + CO_2(g) = M_2CO_2(\ell)$$

and giving the overall reaction

$$CO_2(g) + C(s) = 2CO(g)$$

This mechanism assumes the existence of two different types of sites
on the carbon surface at which the anodic and cathodic reactions
separately occur. An alternate scheme, proposed by Wagner [106],
considers the simultaneous electrochemical reactions,

$$CO_3^{2-} + 2e^- = CO(g) + 2O^{2-} \quad \text{(cathodic reaction)}$$

$$C(s) + O^{2-} = C-O_{ads} + 2e^- \quad \text{(anodic reaction)}$$

followed by the desorption step

$$C-O_{ads} = CO(g)$$

These reactions are feasible under conditions where a continuous electrolyte phase is present. Thus, Lorenz and Janz [107] have shown that CO_2 can be electrochemically reduced to CO in melts containing Li_2CO_3, by the reaction

$$2CO_2(g) + 2e^- = CO(g) + CO_3^{2-}$$

and fuel cells can be operated with carbon electrodes in a molten carbonate electrolyte [89]. Mechanisms of this type are basically oxidation-reduction cycles described in electrochemical terms and include electron-transfer steps as well as oxygen-transfer processes. It is, however, not clear that an electrochemical mechanism can explain the marked catalytic effects of alkali-carbonate additions in the parts per million range, and the concept of a liquid electrolyte phase is clearly inappropriate in the case of transition metal catalysts. However, electrochemical mechanisms represent a novel approach to carbonate-catalyzed carbon gasification processes and suggest the possibility of driving the reactions by the application of an applied potential.

V. CATALYZED C-H$_2$O REACTION

A. Introduction

The endothermic reaction between carbon and water vapor (steam) [26]

$$C(graphite) + H_2O(g) = CO(g) + H_2(g) \quad \Delta H^°_{298.15} = +31.38 \text{ kcal mol}^{-1}$$
$$\Delta G^°_{1200 \text{ K}} = -8.678 \text{ kcal mol}^{-1}$$

is favored by elevated temperature and reduced pressure and, in the absence of a catalyst, occurs sluggishly at temperatures below 1000°C.

The uncatalyzed reaction is generally somewhat faster than the $C-CO_2$ reaction under the same conditions. In a practical system the distribution of products depends on the occurrence of secondary processes, such as the water-gas shift reaction

$$CO + H_2O \rightleftharpoons CO_2 + H_2$$

the Boudouart reaction

$$C + CO_2 \rightleftharpoons 2CO$$

and the hydrogenation reaction

$$C + 2H_2 \rightleftharpoons CH_4$$

In the presence of a catalyst, conditions are usually remote from thermodynamic equilibrium and the distribution of products will depend on the relative rates of the individual catalyzed reactions.

The reaction between steam and carbonaceous materials has, in the past, been of importance in such areas as water-gas manufacture and graphite-moderated nuclear reactor technology. It is of particular relevance at the present time because of the current revival of interest in the steam gasification of coal. Because investment costs for coal-gasification plants would be considerably lowered if the temperature required for the coal-steam and coal-hydrogen reactions could be reduced substantially, much effort is being expended in attempts to catalyze these reactions [108]. For example, in the Battelle Hydrothermal Coal Process [109,110], coal is slurried with a mixture of CaO, NaOH, and water and then heat-treated so that Ca (and to a lesser extent Na) is incorporated into the coal. The Ca serves to render the coal noncaking and to increase its reactivity toward hydrogen and steam. Commercial processes that depend on the catalytic gasification of coal include the Kellogg process [33,35] and the Molten Salt Coal Gasification process of Atomics International [34,111], both of which employ a bath of molten sodium carbonate to catalyze the gasification of coal by air or steam to produce a low or medium BTU fuel gas that is low in sulfur impurities.

Although previously a neglected subject, studies of the catalytic effects of transition metals, noble metals, and alkali and alkaline earth oxides and salts on carbon–steam gasification rates are appearing with increasing frequency, with growing emphasis on the behavior of coal and chars, rather than on graphite, which is used in more basic studies. The behavior of these different kinds of carbonaceous materials is not necessarily similar, in fact the specific rate of the noncatalytic gasification, considered on a weight basis, is often considerably higher for a char than for a pure graphite, although on a unit surface area basis, gasification rates may be comparable [90]. However, catalytic effects are usually more marked with graphite than with chars, because of the lamellar structure of graphite and the ease of penetration of catalyst particles along the basal planes from exposed edges of the crystallites.

B. Catalyzed Steam Gasification of Coals and Chars

The effects of natural impurities in catalyzing the reaction of coals and chars with steam have been reported by several groups of investigators. Linares, Mahajan, and Walker [112] studied the reactivity of chars at 910°C toward wet nitrogen containing 17.5 torr (2.3 kPa) of water pressure at 1 atm (101 kPa) total pressure. The chars were obtained from a variety of parent coals varying from lignite to anthracite. In general, char reactivity decreased with increasing rank or carbon content of the parent coal. In most cases removal of mineral matter by acid leaching resulted in a subsequent decrease in char reactivity. However, with coals of high rank the reverse effect was observed, probably as a result of increases in surface area and porosity on demineralization. Similarly, Otto and Shelef [113] found that acid leaching of a lignite char caused a marked decrease in the rate of steam gasification at 850°C and an increase in the apparent activation energy. Conversely, when 3% by weight of a coal ash, obtained by complete combustion of lignite char, was added to pure graphite a considerable increase in steam gasification rate was obtained.

Elements in the ash, which contained Fe, Ca, Ti, Si, Al, K, and S, were clearly acting as catalysts for the $C-H_2O$ reaction.

In a recent study, Hippo [114] investigated the effect of cation exchange on the subsequent reactivity of lignite chars in steam. Lignite was demineralized by leaching with acid and then ion-exchanged with Na^+, Ca^{2+}, Mg^{2+}, and Fe^{2+}. After charring at 800°C in nitrogen, the ion-exchanged chars showed enhanced steam-gasification rates. For example, at 650°C the gasification rate was more than an order of magnitude greater for the K^+-exchanged char than for the demineralized material. The initial order of reactivity of the cation-exchanged chars for the same concentration of cation (0.3 mol g^{-1} char) was K > Ca, Na, Fe > Mg, although the activity of the Fe-exchanged material decreased with time, presumably as a result of conversion of the iron to inactive oxides. $CaCO_3$ (calcite) was identified in a partially reacted char with the highest Ca^{2+} loading and Na and K were probably also present as the carbonates during the reaction. These catalysts showed no decline in activity with time. In general the reactivity of the chars increased in proportion to the alkali-metal content.

Doping of coal powders with up to 1 at. % of alkaline earths has recently been shown to result in marked increases in the rate of steam gasification of chars derived from the coals, the order of catalytic activity being Ca < Sr < Ba [115]. No significant lowering in the activation energy of the reaction was observed in the presence of the additives, and the study concludes that the catalytic effect is caused by an increase in the density of reaction sites as a result of spreading of the salts over the char surface. However, the mechanism of the catalytic effect was not explored. A marked catalytic activity of calcium in the steam gasification of chars has also been reported by Wilks [116] and Johnson [117]. They observed that ion-exchanged Na^+ and Ca^{2+} increase the reactivity of lignite chars toward both hydrogen and steam. Wilks, Gardner, and Angus [118] have reported that of a wide range of salts investigated, $KHCO_3$ and $NaHCO_3$ were the most active catalysts for steam gasification of coal chars. In this case,

Table 5.   Effect of catalysts on the steam
           gasification of coal char, 940-950°C

Catalyst	Time required for 90% conversion (sec)
5% $KHCO_3$	215
5% $NaHCO_3$	295
5% $ZnCl_2$	370
5% Ag acetate	400
5% $UO_2(NO_3)_2$	535
5% $CoCl_2$	740
No catalyst	890
5% $Pb(NO_3)_2$	1000

Source: From Ref. 118.

5% by weight of added salt was found to result in substantial increases in reaction rate. A summary of the catalytic effects
observed by these authors for a variety of additives is shown in
Table 5. K_2CO_3 alone was also observed to be generally more effective than mixtures of K_2CO_3 with other salts. Surprisingly, lead,
added as the nitrate, exhibited an inhibitory effect on the reaction.

The catalytic effect of iron in promoting the coke-steam reaction is well known and has been studied again recently by Hahn and
Huttinger [119] who vacuum-impregnated coke particles with an iron
sulfate solution to give Fe concentrations in the range 0.05-1.8 wt %.
As expected, the reactivity of the coke toward steam at 890-1000°C
was considerably enhanced in the presence of the Fe catalyst, which
had the effect of lowering the apparent activation energy from
43 kcal mol^{-1} for the uncatalyzed reaction to 19-29 kcal mol^{-1} in
the presence of the Fe impurity. As BET surface areas and porosity
also increased rapidly with burnoff, diffusional effects may have
contributed to the lowering of the apparent activation energy. It
was also found that iron oxide was a less effective catalyst than
metallic iron in this reaction, a conclusion that has also been

reached from studies of the Fe-catalyzed graphite-water vapor
reaction [20].

Some interesting and rather unorthodox conclusions have been
reached by Otto and Shelef [90] who observed that catalysts such as
K_2CO_3 and Ni had little effect on the activation energy for the steam-
gasification reaction on either coal chars or graphite. Thus, addition of 303 ppm of Ni particles to a graphite powder had no effect on
the activation energy (80.5 kcal mol^{-1}, in both cases), but increased
the reaction rate by 2.5 orders of magnitude at temperatures in the
800-1100°C range. With larger amounts of catalyst (e.g., 1.4% Ni) a
decrease in activation energy was observed, but this was attributed
to the influence of mass transport on the kinetics. Although the
majority of investigators have reported a decrease in apparent activation energy associated with the presence of catalysts in carbon-
gasification reactions, Otto and Shelef conclude that the decrease
is merely the result of mass transport or diffusion effects related
to the enhanced porosity of the gasifying carbons. In this study,
catalysis of the steam-gasification reaction was accompanied by an
increase in the preexponential factor, a result that was cited as
evidence for the oxygen-transfer mechanism, involving localized gasification at catalyst sites. An electron-transfer mechanism that would
influence the strength of the C-C bonds in the graphite surface, would
probably result in a change in activation energy, which was not observed in this series of experiments.

In other investigations a decrease in activation energy E in
the presence of an active catalyst has often been observed, and the
change in E and in the corresponding preexponential factor A are
frequently related by a "compensation effect," giving a linear relation of the form

$$\log A = jE + \log k$$

Such a relation has been found by Feistel et al. [120] for steam
gasification of a range of coals, with and without added K_2CO_3

catalyst. The constant j is related in the above expression to the
isokinetic temperature T_s at which all reactions proceed at the same
rate, k,

$$T_s = \frac{1}{Rj}$$

For the series of coals studied by these authors, the value of T_s was
approximately 1150°C.

Kayembe and Pulsifer [121] studied the catalytic effects of
various salts on the kinetics of gasification of coal chars by steam
between 600 and 850°C at atmospheric pressure. The observed order of
activity was K_2CO_3 > Na_2CO_3 > Li_2CO_3 > KCl > NaCl > CuO. Fe_2O_3 and,
in contrast to the results of Hippo [114], CaO were totally ineffec-
tive. Addition of 10% K_2CO_3 to the char lowered the apparent activation energy of the reaction from 254.2 to 144.5 kJ mol^{-1} (61–34.5 kcal
mol^{-1}) and preexponential factor from 2.41 x 10^8 to 1.32 x 10^4 sec^{-1}.
The stoichiometry of the reaction followed the equation

$$C + 2H_2O = CO_2 + 2H_2$$

in the presence of catalysts, regardless of catalyst type, catalyst
concentration, or reaction temperature. In the absence of a catalyst
the product gases were H_2 (52–66%), CO (30–40%), CO_2 (5–9%), and CH_4
(1–2%), the amounts of H_2 and CO_2 decreasing and the amount of CO in
the product gases increasing with increasing temperature. The variation of specific rate constant at 650°C for the various catalysts
used is shown in Table 6.

A large number of metals, oxides, and salts have been investigated for catalytic activity in steam-coal gasification as part of
the Bureau of Mines (DOE) Synthane Process program for converting
coal to synthetic fuel gas [122]. A high-volatile bituminous coal
was pretreated at 450°C with a steam-air mixture to prevent caking,
and the treated coal (which contained 10.6 wt % ash) was then crushed
and sieved to 20-60 mesh. Five wt % of the powdered catalytic additive

Table 6.   Effect of catalysts on the steam
           gasification of coal char [121]

Catalyst (10% by weight)	Rate constant, k (hr^{-1}) at 650°C
CaO	2.438×10^{-3}
Fe_2O_3	3.640×10^{-3}
None	4.250×10^{-3}
CuO	1.014×10^{-2}
NaCl	4.849×10^{-2}
KCl	1.116×10^{-1}
Li_2CO_3	1.670×10^{-1}
Na_2CO_3	3.154×10^{-1}
K_2CO_3	5.388×10^{-1}

Source: From Ref. 121.

was then mixed with the coal and the gasification was carried out at
650–950°C in a tubular fixed-bed reactor using a steam-saturated
nitrogen carrier gas at 300 psig. The steam-gasification reaction
resulted in the formation of both gaseous (H_2, CO, CO_2, CH_4, and
higher hydrocarbons) and liquid (mostly H_2O) products. Typical
gaseous product formation rates and overall gasification rates at
850°C for a 4-hr reaction time are listed in Table 7 for a number of
additives. The results indicate that alkali-metal compounds and the
oxides of Fe, Ca, Mg, and Zn significantly increased the rate of coal
gasification. In contrast to the results obtained by Wilks, Gardner,
and Angus (Table 5), lead oxide was found to be an active catalyst.
Methane formation was increased considerably in the presence of
nickel and significantly by Li_2CO_3, Pb_3O_4, and MgO. H_2 and CO formation was increased most by K_2CO_3, KCl, and Li_2CO_3. It may be noted,
however, that even with the most active catalyst (K_2CO_3) the gasification rate was increased by less than a factor of 2 over that of
the original coal. Separate tests in the Synthane pilot plant

Table 7. Effect of catalysts on the steam gasification of
          bituminous coal, 10 g coal + 0.5 g catalyst 850°C

Catalyst	Total dry gas production rate (N_2 free basis) ($ml\ hr^{-1}\ g^{-1}$) coal	Carbon gasification rate ($g\ hr^{-1}\ g^{-1}$) coal
No catalyst	297	81×10^{-3}
$Co(OH)_2$	345	88×10^{-3}
Ni	366	96×10^{-3}
V_2O_5	367	96×10^{-3}
NiO	373	97×10^{-3}
MnO_2	375	98×10^{-3}
Fe_3O_4	380	100×10^{-3}
MgO	384	100×10^{-3}
BaO	395	104×10^{-3}
PbO_2	400	103×10^{-3}
Pb_3O_4	400	105×10^{-3}
Li_2CO_3	439	113×10^{-3}
ZnO	440	112×10^{-3}
K_2CO_3	578	144×10^{-3}

Source: From Ref. 122.

gasifier at 40 atm pressure and temperatures in the range of 900–945°C
showed that the addition of 5% dolomite or hydrated lime markedly increased gasification rates and product gas yield ($CO + H_2 + CH_4$).

Combinations of K_2CO_3 and a nickel methanation catalyst were
used by Willson et al. [123] to enhance the formation of high Btu gas
from coal by reaction with steam at 650°C and 2 atm pressure. The
alkali-metal carbonate was effective in promoting the steam-coal gasification reaction, whereas the main function of the Ni was to promote
methanation of the carbon oxides and hydrocracking of the liquid
products.

C. Catalyzed Reaction of Carbon and Graphite with Water Vapor

The most frequently studied catalysts for the carbon-H_2O reaction have been the salts (especially the carbonates) of the alkali metals and alkaline earths, iron, cobalt, and nickel and the platinum group metals of group VIII. A few other elements have also been reported to be active catalysts, for example, vanadium has been associated with the occurrence of pitting attack of impure graphite by steam [124,125]; oxides of copper, manganese, zirconium, bismuth, antimony, lead, molybdenum, titanium, chromium, and boron have also been found to be moderately active [122]. Less is known about the effects of catalysts in this case than in the other gasification reactions of carbon. For example, although it has been known for many years that the carbonates of the alkali metals are potent catalysts for the steam gasification of carbon, few attempts have been made to explain this activity.

A recent thermogravimetric study by McKee and Chatterji [126] has suggested a possible mechanism for the catalytic effect of alkali-metal carbonates in the graphite-water vapor reaction. The carbonates of Li, Na, and K (1%, by weight) were mixed with graphite powder, and the rates of gasification in water-saturated helium (P_{H_2O} = 23 torr, 3.1 kPa) were measured at a series of temperatures between 700 and 1100°C. As shown by the Arrhenius plots in Figure 24, the order of catalytic activity for the three salts was $Li_2CO_3 \gg K_2CO_3 > Na_2CO_3$. The apparent activation energies were 87.6 kcal mol^{-1} for pure graphite, 36.7 kcal mol^{-1} for graphite-1% Li_2CO_3, 52.2 kcal mol^{-1} for graphite-1% K_2CO_3, and 53.8 kcal mol^{-1} for graphite-1% Na_2CO_3. Simultaneous TGA-DTA measurements with mixtures of equal weights of graphite powder and Na_2CO_3 (Figure 25) in an inert atmosphere (dry helium) showed that whereas weight loss of the pure carbonate was slight above 1000°C as a result of the dissociation reaction,

$$Na_2CO_3 = Na_2O + CO_2$$

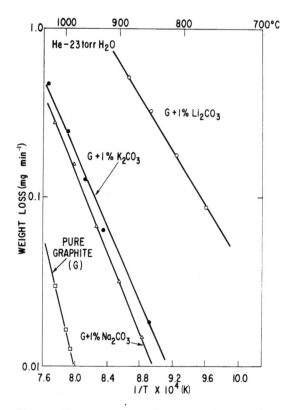

Figure 24. Catalytic effects of 1% Li, Na, and K carbonates on the reaction of graphite with water vapor. Weight loss rates in wet He vs. 1/T K [126].

when the salt was mixed with graphite, evolution of sodium vapor and a rapid loss in weight took place above 900°C because of the reduction reaction,

$$Na_2CO_3 + 2C = 2Na + 3CO$$

In the presence of water vapor, free alkali metal would be rapidly hydrolyzed to NaOH, which however reacts readily with CO at temperatures as low as 500°C to reform the carbonate by

$$2NaOH + CO = Na_2CO_3 + H_2$$

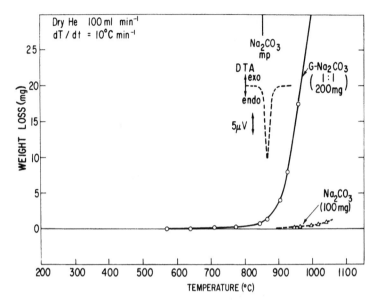

Figure 25. TGA-DTA curves for Na₂CO₃ and a 1:1 graphite-Na₂CO₃ mixture in dry He [126].

As LiOH and KOH would also be unstable in atmospheres containing CO, the alkali-metal hydroxides probably act as unstable intermediate species in the carbonate-catalyzed graphite-H_2O reaction, according to the sequence:

$$M_2CO_3(\ell) + 2C(s) = 2M(g) + 3CO(g) \tag{4}$$

$$2M(g) + 2H_2O(s) = 2MOH(\ell) + H_2(g) \tag{5}$$

$$2MOH(\ell) + CO(g) = M_2CO_3(\ell) + H_2(g) \tag{6}$$

giving the overall reaction

$$C(s) + H_2O(g) = CO(g) + H_2(g)$$

The first of these reactions, the most likely rate-determining step in the sequence, proceeds at a rapid rate at temperatures above the melting point of the carbonate, but the driving force for this

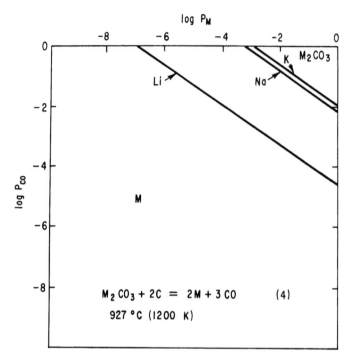

Figure 26. Equilibrium values of P_{CO} and P_M for reaction (4) at 927°C (1200 K) for M = Li, Na, and K. Stability regions of alkali metal, M, and alkali-metal carbonate, M_2CO_3, are indicated [126].

reaction depends on the partial pressures of CO and alkali metal in the vicinity of a reaction site, according to the relation,

$$\Delta G = \Delta G° + RT \ln P_{CO}^3 \cdot P_M^2$$

A plot of the equilibrium pressures of CO and alkali metal resulting from reaction (4) for the Li, Na, and K species at 927°C (1200 K) is shown in Figure 26. For a value of $P_{CO} = 0.1$ atm (10.1 kPa) in the vicinity of the carbon surface (a typical value during the operation of a hypothetical reactor with $P_{H_2O} = 1$ atm), the equilibrium partial pressures of Li, Na, and K in the reaction environment would be approximately 10^{-5}, 10^{-2}, and 10^{-1} atm (1.01×10^{-3}, 1.01, and 10.1 kPa), respectively, as a result of the

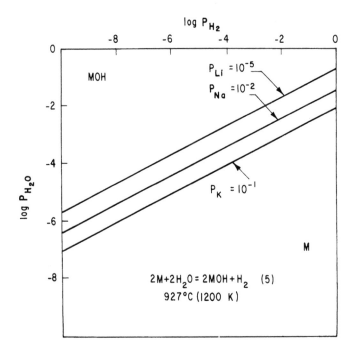

Figure 27. Equilibrium values of P_{H_2O} and P_{H_2} for reaction (5) at 927°C (1200 K) for $P_{Li} = 10^{-5}$ atm (1.01 x 10^{-3} kPa), $P_{Na} = 10^{-2}$ atm (1.01 kPa), and $P_K = 10^{-1}$ atm (10.1 kPa). Stability regions of alkali metal, M, and alkali-metal hydroxide, MOH, are indicated [126].

occurrence of reaction (4). As Na and K are in the gaseous state at this temperature, these alkali metals will tend to vaporize from the carbon surface, whereas Li would tend to remain in the liquid state in contact with the carbon. At 600°C the equilibrium pressures of the alkali metals would be about 2 orders of magnitude less than at 927°C.

The driving force for the hydrolysis reaction (5) would similarly depend on the partial pressures of alkali metal, water vapor, and hydrogen in the vicinity of the carbon surface. The equilibrium stability regions for the Li, Na, and K hydroxides and the corresponding metals calculated at 927°C (1200 K) for reaction (5), are shown in Figure 27, as functions of the values of P_{H_2O} and P_{H_2} in

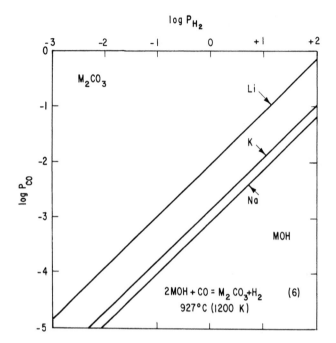

Figure 28. Equilibrium values of P_{CO} and P_{H_2} for reaction (6) at 927°C (1200 K) for M = Li, Na, and K, showing stability regions of alkali-metal hydroxide, MOH, and carbonate, M_2CO_3 [126].

the gas phase. It is evident that over a wide range of water vapor and hydrogen partial pressures the alkali-metal hydroxide will be stable at equilibrium. However, the conversion of hydroxide back to carbonate by reaction (6) also has a strong driving force at this temperature, and M_2CO_3 will be the stable phase for a wide range of P_{CO} and P_{H_2} in the system, as shown in Figure 28.

The sequence of reactions, (4) → (5) → (6) → (4), appears therefore to offer a feasible mechanism for the observed catalytic effects of alkali-metal carbonates in the C-H_2O reaction. The slowest, rate-determining step is probably (4) and the overall gasification rate will be dependent on the salt-carbon interfacial contact area, which is rapidly increased as the salt melts and wets the carbon surface. For this reason, Li_2CO_3, with the lowest melting point and liquid density of the series of carbonates, should exhibit the greatest

catalytic activity. Loss of alkali metal by vaporization will also
be least marked in the case of Li. In general, an increase in the
ambient partial pressure of CO at the carbon surface will decrease
both the driving force for reaction (4) and the overall gasification
rate.

A similar mechanism has recently been proposed by Veraa and Bell
[127] to account for the catalytic effects of alkali-metal salts on
the steam gasification of char obtained from a subbituminous coal.
In the temperature range 700-900°C, K_2CO_3 was the most active additive whereas the alkali-metal chlorides showed an inhibiting effect
in the early stages of the gasification reaction.

Although there have been reports that lime (CaO) or dolomite
($CaCO_3 \cdot MgCO_3$) added to coal can substantially increase the rate of
gasification in steam [122] and also that BaO is an effective catalyst
for the graphite-steam reaction at 1000°C [124], little is known concerning the mode of action of these additives. Everett et al. [125]
found that the addition of 0.05% Ba or Sr salts increased the reactivity of graphite powder toward water vapor at 850°C by a factor
of 1000 in the case of Ba and 130 in the case of Sr. In a recent
thermogravimetric study the author [39] studied the reaction between
water and graphite in the presence of the alkaline earth carbonates
and obtained some clues as to the possible mechanism of the catalytic
effect. A high-purity graphite powder was mixed with 1% by weight of
the salts, and gasification rates in a stream of helium saturated with
water vapor (P_{H_2O} = 23 torr, 3.1 kPa) were measured at a series of
temperatures in the range 800-1100°C. As shown by the Arrhenius plots
in Figure 29, the addition of $BaCO_3$ gave gasification rates about 5
times faster than $SrCO_3$, whereas $CaCO_3$ and $MgCO_3$ showed only a small
catalytic effect. The halides and sulfates of the alkaline earths
were generally much less active catalysts than the carbonates. The
gasification rates and apparent activation energies for the various
carbonate additives are summarized in Table 8. Whereas $BaCO_3$ and
$SrCO_3$ are stable at high temperatures, $MgCO_3$ decomposes to MgO at
500-550°C and $CaCO_3$ to CaO at 700-850°C. Neither MgO nor CaO are

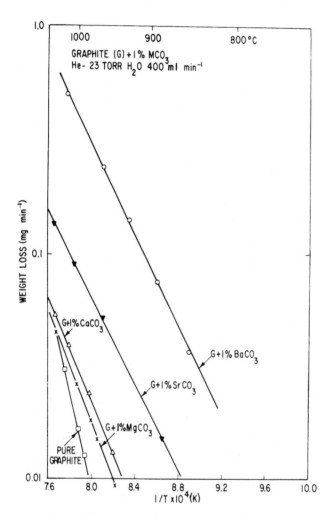

Figure 29.   Catalytic effect of alkaline earth carbonates on the
reaction of graphite with water vapor.   Weight loss rates in wet He
vs. 1/T K [39].

Table 8. Catalytic effect of alkaline earth carbonates
          in the graphite-water vapor reaction

Catalyst	Gasification rate at 1000°C (mg min^{-1})	Apparent activation energy (kcal mol^{-1})
None	0.017	87.6
1% $MgCO_3$	0.025	57.4
1% $CaCO_3$	0.030	53.5
1% $SrCO_3$	0.083	43.3
1% $BaCO_3$	0.400	40.2

Source: From Ref. 39.

reduced by carbon at temperatures below 1000°C nor do these oxides interact appreciably with water vapor at elevated temperatures. Calcium shows some tendency to form a superoxide CaO_2 at low temperatures so it is possible that dissociative chemisorption of H_2O may occur on CaO to some extent at temperatures in the gasification range, thereby leading to enhanced gasification rates in the presence of carbon. Although $SrCO_3$ and $BaCO_3$ are stable in an inert atmosphere to temperatures exceeding 1000°C, they are readily reduced to oxide by reaction with carbon at temperatures of 850°C and above, as shown by the thermograms in Figure 30 for graphite-$BaCO_3$ and graphite-$SrCO_3$ mixtures heated in dry helium. The oxides of the alkaline earth metals are more stable than those of the alkali metals so that formation of free metal is unlikely in the former case. In the presence of water vapor, the alkaline earth oxides are readily hydrated to the corresponding hydroxides at temperatures in the 300-700°C range, as shown by the thermogram in Figure 31 for BaO heated in water-saturated helium. At higher temperatures the hydroxides tend to dehydrate to the oxides, but in the presence of CO, conversion of the hydroxide back to carbonate takes place at temperatures of 850°C and above, as shown by the thermogram in Figure 32 for $Ba(OH)_2$ heated in helium containing 5% CO. Similar reactions occur in the strontium series of compounds.

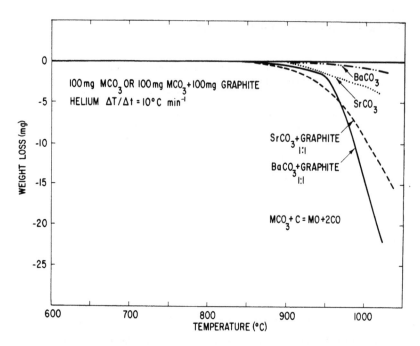

Figure 30. Thermogravimetric (TGA) curves for $BaCO_3$, $SrCO_3$, and 1:1 graphite-$BaCO_3$ and graphite-$SrCO_3$ mixtures on heating in dry He [39].

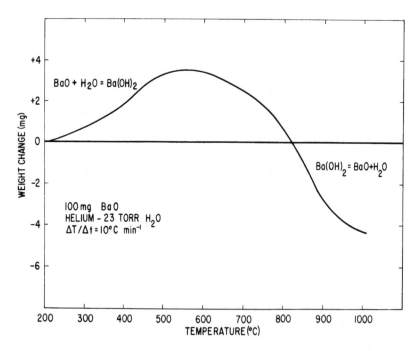

Figure 31. Thermogravimetric (TGA) curve for BaO (100 mg) on heating in wet He. Weight changes vs. temperature [39].

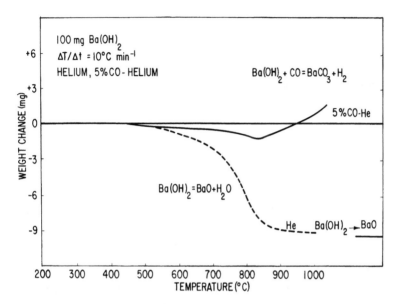

Figure 32. Thermogravimetric (TGA) curves for $Ba(OH)_2$ (100 mg) on heating in dry He and 5% CO-He. Weight changes vs. temperature [39].

It appears likely, therefore, that the most probable oxidation-reduction cycle involved in the $BaCO_3$ or $SrCO_3$-catalyzed graphite-H_2O reaction is the sequence of reaction steps,

$$MCO_3 + C = MO + 2CO \tag{7}$$

$$MO + H_2O = M(OH)_2 \tag{8}$$

$$M(OH)_2 + CO = MCO_3 + H_2 \tag{9}$$

giving the overall reaction

$$C + H_2O = CO + H_2$$

In this sequence, reaction (7) is probably the rate-determining step and reactions (8) and (9) are both rapid at temperatures above 800°C. The hydroxide formed by reaction (8) may have only a transitory existence, and this species probably acts as an unstable intermediate at temperatures in the range 800-1000°C. Both BaO and SrO react with

ambient oxygen at low temperatures to form peroxides, which are themselves very reactive towards carbon in the presence of water vapor.
When the gasification reaction is catalyzed by the alkaline earth
oxides containing peroxides, additional steps are possible, for
example:

$$BaO_2 + C = BaO + CO$$
$$BaO + H_2O = Ba(OH)_2$$

etc. The occurrence of these reactions at temperatures in the catalytic gasification range has been demonstrated by thermol-analysis
techniques [39].

The metals of group VIII are well known to be active catalysts
for the carbon-H_2O reaction but relative activities have been ranked
differently by various investigators. According to Watanabe [128],
at 800°C and 1300 torr (173 kPa) steam pressure, the order of activity
of the group VIII metals is Ru > Rh > Ir > Os > Pd > Co > Ni > Fe.
Otto and Shelef, however, have reported [129] that in the catalyzed
steam gasification of graphite at 850°C the noble metals fall in the
sequence Rh > Ru > Pd > Pt, gasification rates decreasing by about an
order of magnitude from one metal to the next in the series. They
have also observed that for a given metal at a concentration of 1% by
weight, the catalytic activity was about the same whether the metal
was introduced by impregnation from aqueous solution or intercalated
between the graphite basal planes ("graphimets"). Rhodium was also
found by Tomita et al. [130] to be a very active catalyst, and at
650°C a carbon impregnated with 4.8 wt % Rh gave the same conversion
rate in hydrogen as a pure carbon at 1000°C. Rewick, Wentrcek, and
Wise [131] examined the effect of impregnated Pt, Ru, Fe, Co, and Ni
on the kinetics of steam gasification of a thermal carbon black
(Sterling FT) in the temperature range 700-900°C. In this work Pt
was the most active catalyst, as shown in Table 9, which shows the
rates of formation of gaseous products as a function of temperature
and metal additive. In the case of Pt, the overall gasification

Table 9.   Kinetics of the reaction of water vapor (2.4 vol %) with
           Sterling FT carbon containing different metal catalysts

Catalyst	Loading (wt %)	Net gas formation rate $(cm^3 min^{-1} g^{-1}$ carbon)				
		702°C	752°C	802°C	852°C	902°C
None	—	0	0.20	0.22	0.65	1.9
Pt	5	—	~	4.6	18	38
Pt	0.8	0.37	1.1	2.8	7.0	15
Ru	0.8	0.41	0.80	2.1	5.3	11
Ni	0.8	0.10	0.43	1.1	2.2	3.1
Co	0.8	—	—	0.22	0.97	1.6
Fe	0.8	—	—	0	0.09	0.26

Source: Reproduced from Ref. 131, R. T. Rewick, P. R. Wentrcek, and
H. Wise, *Fuel*, 1974, 53, p. 274 by permission of the publishers, IPC
Business Press Ltd. ©.

rate increased with increasing metal content, but when the specific

gasification rates (per unit surface area of Pt) were compared

(Table 10) the sample with the lower Pt content was found to exhibit

the highest specific activity.   This result illustrates again that

the degree of dispersion of a metal catalyst on the carbon substrate

is an important factor in determining catalytic activity.

Table 10.   Specific gasification rates for the catalyzed carbon[a]-
            steam reaction

Catalyst	Wt %	Rate of total gas production[b] $(cm^3 min^{-1} m^{-1}$ Pt)		
		802°C	852°C	902°C
Pt	5	116	456	962
Pt	0.8	651	1630	3490

[a] Sterling FT.
[b] 50 vol % CO, 50 vol % H_2.

Source: Reproduced from Ref. 131, R. T. Rewick, P. R. Wentrcek, and
H. Wise, *Fuel*, 1974, 53, p. 274 by permission of the publishers, IPC
Business Press Ltd. ©.

Tamai et al. [101] have recently studied the catalytic behavior of the group VIII metals in the steam gasification of carbon in some detail. Steam-activated granular carbon was impregnated with aqueous solutions of the metal chlorides then reduced in H_2 at 300-500°C to give a 5 wt % dispersion of the metals. Gasification was accomplished in a thermobalance at a heating rate of 200°C an hour, using steam-saturated helium. Effluent gas analysis was performed by gas chromatography. Figure 33 shows the reactivity profiles as a function of temperature for the metal catalysts investigated.

The metals were divided into three groups according to their behavior as catalysts in the gasification reaction; as follows. (1) Ni, Ru, Rh, and Os showed a reactivity peak at low temperatures (500-700°C) and a fairly high general level of catalytic activity. (2) Ir and Pt showed a high activity at high temperature (> 700°C) but no low-temperature peak.   (3) Fe, Co, showed modest activity at high temperatures, whereas Pd was virtually inert. By contrast, most others have found that Ni is generally less active than Fe in this reaction.

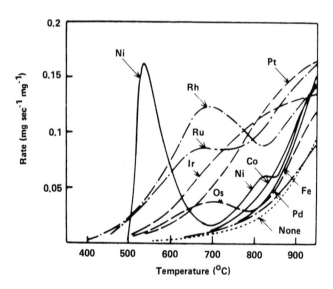

Figure 33. Catalytic steam gasification of active carbon. Rates vs. temperature for various catalysts [101].

The ratio CO_2/CO in the product was found to vary with temperature and the metal catalyst present. At 800°C Fe gave the highest
value for this ratio (1.3) and Ir the lowest (0.08), and in general
the higher the catalytic activity the lower the CO_2/CO ratio, which
approached the theoretical value calculated for the equilibria

$$C + H_2O = CO + H_2$$
$$CO + H_2O = CO_2 + H_2$$

only for conversions of 50% or more. No explanation was offered by
the authors for the low-temperature reactivity shown in Figure 33,
although similar effects were observed for the $C-CO_2$ and $C-H_2$ reactions on the same impregnated carbon samples (Figures 21 and 40).

In a recent investigation, the present author [20] studied the
catalytic effects of a number of metals on the reactivity of graphite
toward water vapor in both oxidizing and reducing atmospheres. Fe,
Ni, and Co were active catalysts for the reaction in the temperature
range 600-1000°C but only when the metal was kept in the reduced
state by means of hydrogen added to the gas phase. The effect of Fe
concentration on the reactivity of graphite in wet hydrogen is shown
in Figure 34. The results of thermogravimetric experiments carried out
at 930°C with graphite impregnated with 1 wt % Fe are shown in Figure
35 for three different gaseous atmospheres. Initially, using nitrogen
saturated at room temperature with water vapor (23 torr H_2O, 3.1 kPa),
no weight loss of the graphite was detected for a period of 30 min, at
which point the wet-nitrogen gas stream was replaced by wet hydrogen.
A rapid gasification of the graphite immediately occurred which subsequently decreased to a much lower rate when the water vapor was removed
from the hydrogen stream. The graphite-water reaction thus appeared
to be considerably more rapid than the graphite-hydrogen reaction at
this temperature. In separate experiments it was found that graphite
containing 1% Fe reacted at a given rate with water vapor at temperatures about 300°C lower in the presence of hydrogen than when nitrogen
was used as the carrier gas. Similar effects were observed with Ni

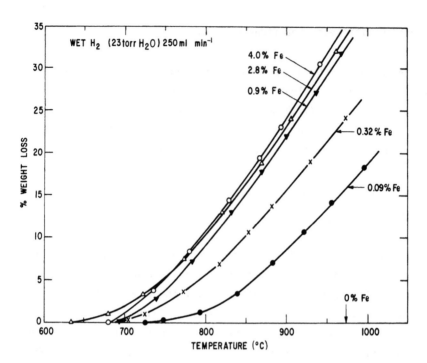

Figure 34. Effect of Fe on the reaction of graphite powder with water vapor. Percent weight loss vs. temperature [20].

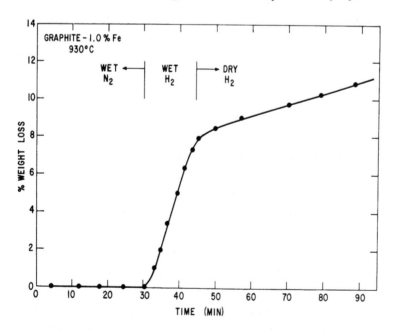

Figure 35. Reaction of graphite-1% Fe with wet (23 torr H_2O) and dry H_2. Percent weight loss vs. time at 930°C. Gas flow rate = 250 ml min^{-1} [20].

and Co, which were rather less active than Fe as catalysts for this
reaction. In all three cases the metal itself appeared to be the
catalytically active entity, whereas the oxides were inactive. Mo
and V also exhibited some weak activity for the graphite-water reaction under reducing conditions, but Cu, Zn, Mn, Cr, Cd, Pb, and Ag
were inactive, even in the presence of hydrogen.

Hahn et al. [132] have suggested that the reactivity of carbon
may be enhanced by dissolution in the metal phase — a process which is
rapid at the temperature of the steam-gasification reaction. However,
as it is known that metals such as iron chemisorb H_2O dissociatively,
even at low temperatures [133], it appears likely that dissociation
of water occurs initially at the metal catalyst surface, leading to
chemisorbed O and H atoms.

$$3Fe + H_2O = Fe-O_{ads} + 2(Fe-H_{ads})$$

This may be the rate-determining step in the overall gasification
reaction, and the higher the affinity of the metal for oxygen, the
more rapid will be this oxygen-transfer step and the greater the
catalytic activity of the metal. Predictably, the catalytic activity
should therefore decrease in the order Fe > Co > Ni, as found experimentally. Metals, such as Cu, Ag, and Pb, do not promote the dissociation of H_2O and are inactive. Other metals, such as Cr and Mn, have
such a high affinity for oxygen that the subsequent reaction with the
graphite substrate

$$Fe-O_{ads} + C = Fe + C-O_{ads}$$

and desorption of CO

$$C-O_{ads} = CO(g)$$

will not take place, and these metals will also be inactive. The
minimum H_2/H_2O ratios needed to convert the lowest metallic oxides
to the metallic state at 1000°C can be calculated from free-energy

data to be: Ni 1×10^{-2}, Co 5×10^{-2}, Fe 3.0, Cr 5×10^3. Thus, for
the wet-hydrogen environment used in the author's experiments [20]
$H_2/H_2O = 32$; Fe, Co, and Ni will exist in the metallic state at equilibrium, whereas Cr will remain in the oxidized form. In wet nitrogen,
all four metals would be present as oxides. Thus it is reasonable to
conclude that metallic particles are the active phases responsible for
catalyzing the graphite-H_2O reaction at temperatures in the vicinity
of 1000°C. Also, during the catalyzed reaction the particles of Fe,
Co, and Ni were observed to move rapidly on the basal-plane surfaces
of graphite single crystals, the channels produced by the smallest
particles being oriented predominately in the $< 11\bar{2}0 >$ directions.
Catalyst particles larger than 10 μm showed more sluggish motion and
tended to form curved channels. These features of the behavior of
Fe particles on graphite surfaces have been confirmed recently [77]
by electron microscopy.

VI. CATALYZED C-H_2 REACTION

A. Introduction

The reaction between carbon and hydrogen [26],

$$C + 2H_2 = CH_4 \quad \Delta H°_{298.15} = -17.895 \text{ kcal mol}^{-1}$$
$$\Delta G°_{1200 \text{ K}} = +9.887 \text{ kcal mol}^{-1}$$

is the least favorable, thermodynamically, of the gasification reactions of carbon and, in the absence of a catalyst, requires even
higher temperatures and pressures than the endothermic C-H_2O reaction.
At 1000°C and 1 atm pressure, the equilibrium concentration of CH_4 in
the hydrogenation reaction is only about 1%, and therefore the reaction must be carried out at elevated pressures if a reasonable yield
of methane is to be attained. The equilibrium yield of CH_4 increases
with decreasing temperature, hence the need for catalysts to improve
the kinetics at lower temperatures. The reverse reaction, involving

the dissociation of methane, can be carried out at temperatures as
low as 650°C on the surfaces of catalytically active metals [134].

B. Catalyzed Hydrogenation of Coal and Coke

Motivated by the need to improve the efficiency and to lower the
operating temperature of coal-gasification processes, interest in
catalysts for the hydrogenation of coal (hydrogasification) has pro-
liferated since the mid-seventies. Materials that catalyze the reaction between coal and hydrogen can be divided into three classes:

1. Salts and oxides of the alkali and alkaline earth metals
2. Fe, Co, Ni, and the noble metals of group VIII
3. Ammonium molybdate, sulfides of Mo and W, and the chlorides
    of Zn, Al, and Sn

Catalysts of the first and second groups promote the formation
of gaseous hydrocarbon products, whereas those of the third group are
effective in increasing the yields of liquid hydrocarbons. The latter
will not be discussed further as this topic is outside the scope of
the present review [135].

A number of studies have been concerned with the catalytic
effects of natural coal minerals and impurities on the reactivity
of coals in hydrogen.

Thus Mukherjee and Chowdhury [136] found a clear correlation
between the Fe and Ti contents of coal ash and the rate of gasification of the corresponding coal in H_2.

Tomita et al. [137] demonstrated that removal of mineral matter
from chars by acid washing generally resulted in a decrease in reactivity of the char toward H_2. However, the demineralization process
can result in profound changes in surface area and porosity, which
can alter the reactivity-temperature profile of the char. The same
investigators [32] have shown that the coal minerals siderite ($FeCO_3$)
and pyrite (FeS_2) can have marked catalytic effects in the hydrogasification reaction, both iron minerals being reduced to metal during

the reaction. On the other hand, the minerals illite, gypsum, rutile, kaolinite, and especially calcite had an inhibitory effect on the hydrogasification reaction. This suggests that the latter minerals inhibit the reaction by providing sites for H-atom recombination, thereby reducing the rate of char gasification.

Johnson [117] showed that ion-exchanged cations, particularly Na^+ and Ca^{2+}, present in lignites on carboxyl functional groups, can catalyze char-hydrogen reactions. Figure 36 shows the effects of added Na^+ and Ca^{2+} on the reactivity of a number of lignite chars. The ordinate f_L/f_L^o is the ratio of the reactivity of Na^+- and Ca^{2+}-exchanged lignite to that of the acid-treated demineralized material.

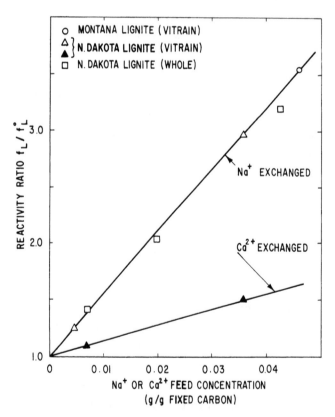

Figure 36. Effect of ion exchange on the reactivity of lignite char in H_2 at 926°C and 35 atm [117].

The catalytic effect was found to be greater in steam-H_2 than in H_2 alone.

Treatment of coals with lime (CaO) according to the Battelle Hydrothermal Coal Process has been claimed to increase the rates of hydrogasification by as much as 40 times over that of raw coal [109]. Thus Chauhan et al. [110] have reported that incorporation of CaO + NaOH, or CaO alone, into a coal increased the reactivity of the coal toward H_2 at 800–1000°C. A comparison of typical product yields and reactivities for a treated coal with added CaO + NaOH and a preoxidized coal without additive is shown in Table 11.

The CaO present in the treated coal also served to reduce the S content of the product gases, as a result of the formation of CaS. However, no decline in activity of the CaO-treated coal was observed up to a conversion of 75%.

Gardner et al. [138] studied the reaction between hydrogen and coal char (hydrogasification) in the presence of various additives. K_2CO_3 and $KHCO_3$ were generally more effective catalysts than $ZnCl_2$.

Table 11. Catalytic hydrogenation of coal treated with CaO + NaOH

	Treated coal (CaO + NaOH)	Preoxidized coal (no additive)
Total pressure psia	265	265
P_{H_2} in feed gas psia	265	265
Temp. range °C	755–1045	850–930
Reaction time, min	13	12
(wt % C Conversion)		
CH_4	36.6	21.2
C_2H_4	2.0	0.3
C_2H_6	11.1	2.0
Tar + oil	5.2	3.0
Oxides	4.3	3.5
Total	59.2	30.0

Source: From Ref. 110.

Thus, at 950°C the time required to produce a given conversion was roughly halved on addition of 5 wt % $KHCO_3$.

Few studies have been made on the effect of metal catalysts on coal hydrogenation rates. Weber and Bastick [139] have observed that for nickel deposited on a coke that had been pyrolysed at 900°C, treatment with hydrogen produced methane at temperatures between 400 and 650°C and above 750°C. The low-temperature evolution was accompanied by an endothermic DTA peak and was apparently due to desorption of CH_4 formed during the coke pyrolysis, whereas CH_4 formed above 750°C gave an exothermic peak and resulted from the Ni-catalyzed hydrogenation reaction. However the low-temperature formation of CH_4 has frequently been observed during the metal-catalyzed hydrogenation of carbon (Section VI.C).

C. Catalyzed Reaction of Carbon and Graphite with Hydrogen

The group VIII metals are well known to be active catalysts for the $C-H_2$ reaction and have often been studied using graphite as the carbonaceous material. These metals are also excellent catalysts for gas-phase hydrogenation and hydrogenolysis reactions, and there appears to be a clear connection between the two functions.

In a recent thermogravimetric study [20], the author examined the catalytic effects of a number of metals on the kinetics of hydrogenation of graphite at temperatures between 600 and 1000°C. High-purity graphite powder was impregnated with aqueous solutions of the acetates of Ni, Co, Cu, Zn, Mn, Cr, Pb, Ag, Cd, and ferric formate. Salts of organic acids were used as these decompose to the respective metal oxides at temperatures below 500°C without contaminating the graphite with residual anions. Graphite powder was also mixed with powdered MoO_3 and V_2O_5. Weight-change measurements at a linearly increasing temperature rate in a stream of dry hydrogen indicated that metallic Fe, Co, and Ni were active catalysts for the gasification reaction at temperatures above 600°C. Mn, Cr, Mo, and V oxides had measureable catalytic activity at temperatures of 900°C and above,

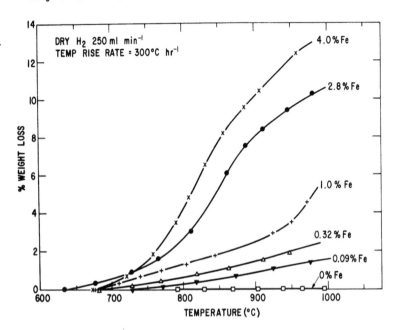

Figure 37. Effect of Fe on the reaction of graphite powder with dry
H_2. Percent weight change vs. temperature [20].

whereas Zn, and Cd oxides and metallic Cu, Pb, and Ag were inactive.
The effect of iron concentration on the temperature and extent of
gasification is shown in Figure 37. Baker and Skiba [19] recently
used CAEM to examine the behavior of silver particles dispersed on
graphite crystals when heated in various gaseous environments. In
dry hydrogen no etch pits or channels were observed. It was concluded that silver is not an active catalyst for the graphite-hydrogen
reaction. The catalytic effects of the noble metals have been the
subject of a number of investigations. For example, Rewick et al.
[131] examined the catalytic effects of Pt, Ru, and Rh in the hydrogasification of several different carbons. At constant H_2 pressure,
the reaction kinetics followed the relation

$$\frac{-dm}{dt} = km$$

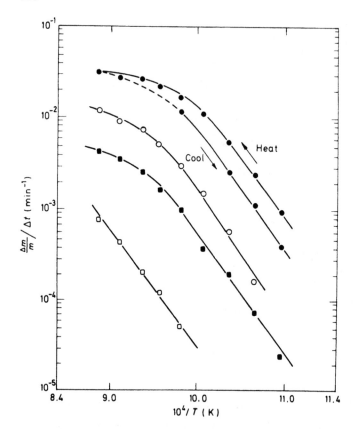

Figure 38.   Gasification rates of Pt-doped carbons in 1 atm H_2 vs.
1/T K.   ● Norit + 5% Pt,   ○ Sterling FT + 0.8% Pt,   ■ Sterling FT + 3%
Pt, □ Graphon + 0.8% Pt.   Reproduced from Ref. 131, R. T. Rewick,
P. R. Wentrcek, and H. Wise, *Fuel*, 1974, 53, p. 274 by permission
of the publishers, IPC Business Press Ltd. ©.

where m is the mass of the carbon and k, the rate constant, was equal
to A exp(-E/RT), CH_4 being the only detectable reaction product.

A plot of k vs. 1/T for the various Pt-doped carbons is shown in
Figure 38.   Although the different samples exhibited marked differences in reactivity, the apparent activation energies were very similar and equal to 55 ± 3 kcal mol^{-1} in every case.   In the linear
regions of the Arrhenius plots, the rate of gasification was proportional to $P_{H_2}^{\frac{1}{2}}$ for one carbon sample (Norit-A) containing 5 wt % Pt
on gasification with different constant values of P_{H_2}.   A comparison
of the behavior of the three noble metals is shown in Figure 39 for

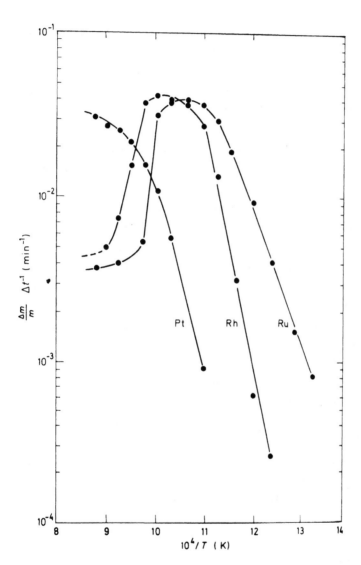

Figure 39. Gasification rates in 1 atm H_2 of Norit carbons containing 5 wt % Pt, Rh, and Ru vs. 1/T K. Reproduced from Ref. 131,
R. T. Rewick, P. R. Wentrcek, and H. Wise, *Fuel*, 1974, 53, p. 274
by permission of the publishers, IPC Business Press Ltd. ©.

the Norit carbon. At 602°C the rate of CH_4 formation was 100 times
as great for Ru and 25 times as great for Rh as for Pt, but a maximum
was observed in the activity of Ru and Rh at higher temperatures.
The occurrence of the maxima was interpreted as due to competition
between two surface processes, one, the continuous formation of a
highly reactive C surface as a result of the gasification reaction
and, the other, thermal annealing or removal of reactive edge atoms
at the higher temperature. However, Marsh and Taylor [140] point out
that the temperatures of maximum rate observed frequently in the gasification of nonporous carbons (e.g., graphite, carbon filaments) occur
at much higher temperatures (1230-1830°C) than those reported by
Rewick et al. (~700°C). Hence, it is likely that the maxima observed
by Rewick et al. with their microporous carbon samples result from
the onset of diffusional control of the kinetics of the gasification
reactions at temperatures in the neighborhood of 700°C.

By measuring the actual surface area of the metallic component
and calculating specific gasification rates per unit area of metal,
Rewick et al. [131] compared the experimental rate of CH_4 production
in the Pt-catalyzed reaction with the net rate of H-atom formation
by dissociation of molecular H_2 on the platinum surface, values for
which are available in the literature [141]. Thus the rate of CH_4
formation on Sterling FT/5% Pt showed the same kinetic dependence on
P_{H_2} and the same activation energy as the rate of H-atom production
on a Pt surface. Table 12 illustrates the close agreement between
the rate of CH_4 formation and the rate of H-atom production. The
authors conclude that the formation of atomic H is the rate-controlling
step in the C-H_2 reaction, the role of the catalyst being to act as a
source of atomic H, which diffuses to the neighboring C sites by surface migration according to the sequence of reaction steps,

$$H_2 + 2M \rightarrow 2(M - H_{ads})$$
$$(M - H_{ads}) + C \rightarrow M + (C - H_{ads})$$
$$(C - H_{ads}) + H_{ads} \rightarrow \cdots \rightarrow CH_4(g)$$

Table 12. Comparison between observed rate of methane and atomic
           hydrogen formation [131]

Temp. °C	H formation [141] (atoms cm^{-2} sec^{-1})	CH_4 formation (molecules cm^{-2} sec^{-1})
652	$3.0 \pm 0.3 \times 10^{14}$	1.2×10^{14}
752	$4.5 \pm 0.7 \times 10^{15}$	3.8×10^{15}

Source: Reproduced from Ref. 131, R. T. Rewick, P. R. Wentrcek,
and H. Wise, *Fuel*, 1974, 53, p. 274 by permission of the publishers,
IPC Business Press Ltd. ©.

The catalytic effects of the transition metals in the C-H_2
reaction can be utilized in the purification of the metals. Thus
sponge iron, chromium, and ferrochromium have been purified from
residual carbon by hydrogenation at 800°C [142].

In the first of a series of investigations, Tomita and Tamai
[143] studied the reaction between several carbons, including graphite,
and H_2 at temperatures up to 1050°C in the presence of 4.8 wt % of the
metals of group VIII, which were impregnated into the carbons as
aqueous solutions of the chlorides. The reaction was studied in
flowing H_2 in a temperature-programmed, quartz-spring thermobalance,
and the gaseous products (mainly methane) were analyzed by a gas
chromatograph. In every case CH_4 was formed in several stages with
a maximum hydrogenation rate at intermediate temperatures in the
400-800°C range. The temperature for maximum gasification rate varied
however with the type of carbon, the metal present and its degree of
dispersion. Pitting and catalytic channeling were also observed in
the vicinity of metal particles after heating in H_2 to 1050°C. The
results were interpreted qualitatively in terms of the dissociative
adsorption of hydrogen at metal sites, followed by the migration of
H atoms to the carbon substrate [144]. The observed maxima in reactivity at different temperatures was probably related to the existence
of surface heterogeneities in the various carbon samples. The relative orders of activity in the lower temperature region was found to
be Rh > Ru > Ir > Pt > Ni >> Pd > Co > Fe. This sequence is essentially the reverse of that found for the CO_2-catalyzed reaction [6]

but is similar to the order of activity of the metals in gas-phase
hydrogenation reactions, such as the catalytic hydrogenolysis of
ethane at low temperatures [145].

In a later paper, Tomita et al. [130] reported that the hydrogenation rate vs. temperature profile for carbons impregnated with
Ni, Pt, and Rh was very dependent on the nature of the carbon. For
example, a Rh-doped active carbon again gave a bimodal pattern of
gasification rates, with maximum CH_4 yields at 600 and 800°C. Rh-
doped carbon black showed similar behavior, whereas Rh-doped graphite
showed only one peak of reactivity at 920°C. Similar results were
observed with an active carbon doped with Ni, which gave peaks in
CH_4 yield at 550 and 950°C and with a Pt catalyst, which gave peaks
at 770 and 1000°C. However, in the absence of a metallic catalyst,
the low-temperature gasification process was not observed, CH_4 formation becoming observable only above 850°C for the uncatalyzed reaction. The presence of two stages in the gasification process with
increasing temperature was attributed by the authors to the presence
of two components in the carbon having different reactivities; for
example, an amorphous phase, which is gasified at low temperatures
by H atoms migrating from metallic sites and a more crystalline
fraction, which reacts only at higher temperature.

In a more recent study, Tamai et al. [101] examined the catalytic effects of the group VIII metals in the carbon-H_2 reaction
between 400 and 1000°C. A steam-activated granular carbon was impregnated with aqueous solutions of the metal chlorides and then
reduced in H_2 at 300–500°C to give 4.8% of metal by weight. Gasification rates were measured in H_2, CO_2, and H_2O vapor in a temperature-
programmed thermobalance. The relative activities of the metals were
about the same for all three gases, Ru, Rh, Ir, and Pt being generally
more active than Fe, Co, and Pd. As shown in Figure 40, for the C-H_2
reaction, Ni, Ru, Rh, and Os exhibited a maximum in the reactivity vs.
temperature plot in the range 500–600°C and a similar pattern of
behavior was observed with CO_2 and H_2O as the reactant gas. No
explanation was offered for this effect.

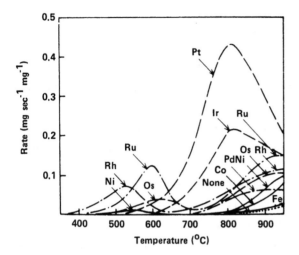

Figure 40. Effect of catalysts on the reactivity of active carbon in H_2. Gasification rates vs. temperature for various metal additions [101].

Weber and Bastick [146] have also observed that in the presence of a nickel catalyst, the hydrogenation of carbons often occurs in two distinct steps with increasing temperature, the formation of methane rising to a maximum rate between 400 and 650°C and then increasing again above 750°C. The reactivity at the lower temperature was observed to be a transitory effect, again probably associated with the presence of a more reactive phase in the carbon that was gasified preferentially at a lower temperature than the bulk of the carbon. However, in a later paper [147] Weber and Bastick suggest that the progressive loss of activity in the low temperature region may be due to deactivation of the catalyst rather than to loss of the most reactive fraction of the carbon sample.

In general terms the catalytic effects of the group VIII metals in the carbon-hydrogen reaction are the most easily understood of the catalyzed gasification reactions of carbon. It appears to be generally agreed that the metal particles dispersed on a carbon substrate act as dissociation centers for molecular H_2 and the resulting H atoms then

migrate to the carbon phase where the carbon π-bonds are readily
hydrogenated. Cleavage of the resulting carbon σ-bonds is then
followed by conversion of the methyl groups into methane, which is
desorbed as a gaseous product. The differences in catalytic behavior
between individual metals of group VIII are less clearly understood,
but the activity in carbon hydrogasification generally follows the
sequence of activity observed in gas-phase hydrogenation or hydrogenolysis reactions.

Thus, Robertson et al. [148] have found that as with hydrogenolysis reactions, the catalytic activity of Ni for the C-H_2
reaction is reduced on alloying with a group IB metal. As shown in
Figure 41, the peak in methane yield at 550-600°C observed with the
Ni catalyst was substantially reduced by alloying with Cu, although
the increase in gasification rate above 700°C was unaffected.

The surface migration of H atoms from a catalytically active
metal site to a carbon substrate has been well documented and is
frequently termed "spill-over" [149,150]. In order to explain the
observed high rates of transport of atomic H, Boudart et al. [144]
have suggested that the metal surface itself is generally contaminated with carbon which provides bridges for the surface transport
of H atoms from the metal surface to the neighboring carbon phase.
An electrochemical explanation of H-atom spillover has recently been
advanced by Keren and Soffer [149]. Molecular H_2 is assumed to be
dissociated at metallic sites into protons and electrons, both of
which migrate onto the carbon surface, proton migration proceeding
by the formation of H bonds bridging the hydroxyl surface groups.
The driving force for the transport of both protons and electrons is
assumed to be the electrochemical potential gradient between the metal
sites and the carbon sites and the amount of hydrogen uptake is determined by the capacity of the electrical double layer at the carbon
surface.

As with the other catalyzed gasification reactions, the reaction
of H_2 with graphite crystals in the presence of metallic catalysts is
often accompanied either by the formation of pits in the vicinity of

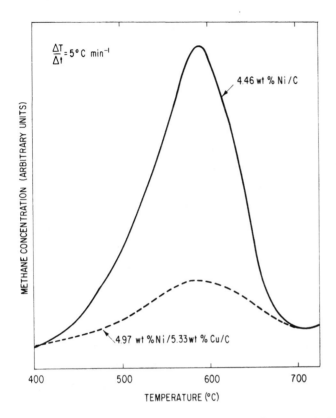

Figure 41. Effect of Ni and Ni–Cu on the reactivity of active carbon in H_2. CH_4 yield vs. temperature [148].

stationary metal particles or by channeling in the basal plane by mobile catalyst particles. The topography of channels produced by Fe, Co, Ni, and Rh particles during the hydrogenation of natural flake graphite at 1050°C has been studied microscopically by Tomita and Tamai [8].

In this study the channels produced by the mobile metallic particles were mostly in random directions on the basal plane, although in some cases straight channels in the preferred $< 11\overline{2}0 >$ direction, i.e., perpendicular to the $< 10\overline{1}0 >$ twin bands, were observed, especially with the smallest particles. Larger particles tended to form curved channels. All channels appeared to start at steps on the

graphite surface, whereas particles on flat regions of the basal
plane were immobile and catalytically inactive. With irregularly
shaped particles the leading edge during channel formation was always
the longer side, supporting Hennig's [45,151] suggestion that the
driving force for catalyst particle mobility is an attractive force
between the particle and edge carbon atoms at the channel tip.

The preferred $< 11\overline{2}0 >$ direction of channel propagation was
rationalized by the authors in terms of the elementary reactions
occurring at labile carbon atoms on the channel sides. In Figure 42
the reaction of H atoms with the adjacent labile carbon atoms on the
"armchair" or $\{11\overline{2}\ell\}$ face is termed a reaction of type A and would be
expected from general energy considerations to occur more readily than
a type B reaction at the "zigzag" or $\{10\overline{1}\ell\}$ face where alternate carbon
atoms have unsaturated valences. The formation of uniformly oriented,
hexagonal etch pits during the uncatalyzed oxidation of graphite crystals is generally assumed to be the result of preferential gasification
of carbon atoms having the labile "armchair" configuration. Recent
calculations by Abrahamson have indicated that the surface energy of
the "armchair" configuration is about 15% greater than that of the
"zigzag" arrangement [152]. In the case of the hydrogenation reaction,
a type A reaction would result in the rapid removal of "armchair"
carbon atoms by a zipperlike process, resulting in the formation of
a $\{10\overline{1}\ell\}$ edge array. A reaction of type C, occurring at the cross
point of two $\{10\overline{1}\ell\}$ faces, would be expected to be relatively infre-
quent.

An idealized channel propagating in the $< 11\overline{2}0 >$ direction is
shown schematically in Figure 43. The sides and tip of the channel
consist of $\{101\ell\}$ faces with exposed carbon atoms in the "zigzag"
configuration. As the channel direction is perpendicular to the
direction of highest reactivity, the ease of occurrence of reactions
of type A results in the elimination of "armchair" sites and the
orientation of the channel in the $< 11\overline{2}0 >$ direction. The further
propagation of such a channel will depend on the occurrence of reactions of type B at the points of contact of the channel sides with

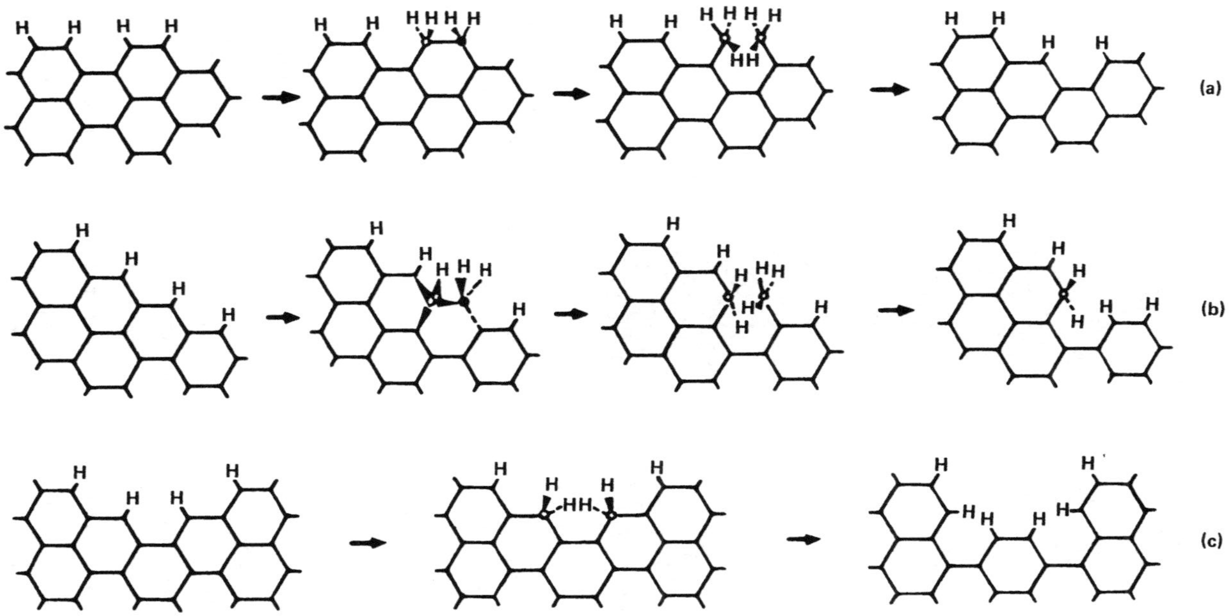

Figure 42. Three possible reaction types in the hydrogenation of graphite: (a) armchair {112ℓ} face, (b) zigzag {101ℓ} face, (c) cross point of two {112ℓ} faces (according to Tomita and Tamai [8]). Reprinted with permission from A. Tomita and Y. Tamai, J. Phys. Chem., 78, 2254 (1974). Copyright by the American Chemical Society.

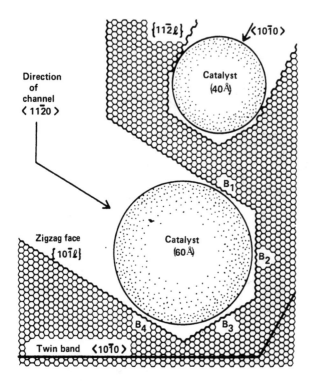

Figure 43. Schematic representation of the propagation of a cata-
lytic channel in the < 11$\bar{2}$0 > and < 10$\bar{1}$0 > directions (according to
Tomita and Tamai [8]). Reprinted with permission from A. Tomita and
Y. Tamai, *J. Phys. Chem.*, 78, 2254 (1974). Copyright by the American
Chemical Society.

the catalyst particle or on reactions of type C at the channel tip.

Reactions B and C therefore determine the overall rate of reaction

and the geometrical pattern of channels. Reaction at sites B_2 and

B_3, followed by rapid reactions of type A would result in rectilinear

motion of the catalyst particle, whereas reaction at B_1 and B_2 or

B_3 and B_4 would result in changes in the channel direction of 120°

that are often observed, especially with larger particles. The

propagation of a straight channel is more difficult with irregularly

shaped catalyst particles, which are generally observed to form

channels of random orientation with frequent changes in direction.

This model represents one of the few attempts to interpret the features of catalytic channeling on an atomistic basis, and the general points should be applicable to other types of gasification reactions.

VII. CONCLUSIONS

During the past decade the diverse catalytic effects of inorganic impurities on the gasification reactions of carbon have continued to provide a fruitful area for basic and applied research and a testing ground for theories both old and new. The subject appears to be growing in industrial importance, especially in connection with the vigorously expanding field of coal gasification and processing, and significant breakthroughs may be confidently predicted in this technology. On the more basic side, the advent of new experimental techniques and detailed consideration of the elementary reactions involved have provided better insight into the mode of action of individual catalysts. Although it is not yet possible to explain all the observed catalytic effects within one all-encompassing, mechanistic framework, on balance specific oxidation-reduction cycles have been conspicuously successful in interpreting the effects of alkali-metal salts, transition metals and oxides, and the noble metals in the various types of carbon gasification reactions. However, many details of the complex catalytic phenomena still remain obscure and await elucidation by carefully designed experimental and theoretical investigations.

ACKNOWLEDGMENTS

The author is indebted to his colleagues D. Chatterji and A. Urquhart for advice and encouragement during the preparation of this manuscript, to R. T. K. Baker for Figures 6, 7, 11, to H. Marsh for Figures 19 and 20, and to the editors of Carbon (Pergamon Press), the Journal of Catalysis (Academic Press), Fuel (IPC Press), and the American Chemical Society for permission to reproduce data and figures.

REFERENCES

1. C. Kröger and Z. Angew, *Z. Angew Chem.*, 52, 129 (1939).

2. P. L. Walker, Jr., F. Rusinko, Jr., and L. G. Austin, in *Advances in Catalysis*, Vol. 11 (D. D. Eley, P. W. Selwood, and P. B. Weisz, Eds.), Academic, New York, 1959, p. 133.

3. P. L. Walker, Jr., M. Shelef, and R. A. Anderson, in *Chemistry and Physics of Carbon*, Vol. 4 (P. L. Walker, Jr., Ed.), Marcel Dekker, New York, 1968, p. 287.

4. J. B. Lewis, in *Modern Aspects of Graphite Technology* (L. C. F. Blackman, Ed.), Academic, New York, 1970, p. 129.

5. D. W. McKee, *Carbon*, 8, 131 (1970).

6. D. W. McKee, *Carbon*, 8, 623 (1970).

7. J. W. Patrick and A. Walker, *Carbon*, 12, 507 (1974).

8. A. Tomita and Y. Tamai, *J. Phys. Chem.*, 78, 2254 (1974).

9. R. R. Adair, E. H. Boult, E. M. Freeman, S. Jasienko, and H. Marsh, *Carbon*, 9, 763 (1971).

10. R. J. M. Griffiths and E. L. Evans, *J. Catal.*, 36, 413 (1975).

11. H. Marsh and B. Rand, *Carbon*, 9, 63 (1971).

12. R. T. K. Baker and P. S. Harris, *J. Sci. Instrum.*, 5, 793 (1972).

13. R. T. K. Baker, *Chem. Eng. Progress*, 73(4), 97 (1977).

14. P. S. Harris, Ph.D. Thesis, University of Surrey, U.K., 1974.

15. R. T. K. Baker, P. S. Harris, and R. B. Thomas, *Surface Sci.*, 46, 311 (1974).

16. R. T. K. Baker, P. S. Harris, D. J. Kemper, and R. J. Waite, *Carbon*, 12, 179 (1974).

17. R. T. K. Baker, R. B. Thomas, and M. Wells, *Carbon*, 13, 141 (1974).

18. R. T. K. Baker, J. A. France, L. Rouse, and R. J. Waite, *J. Catal.*, 41, 22 (1976).

19. R. T. K. Baker and P. Skiba, Jr., *Carbon*, 15, 233 (1977).

20. D. W. McKee, *Carbon*, 12, 453 (1974).

21. D. W. McKee and D. Chatterji, *Carbon*, 13, 381 (1975).

22. A. O. Wist, *Proc. 3rd Int. Conf. Thermal Anal.*, 1, 515 (1971).

23. F. H. Franke and M. Meraikib, *Carbon*, 8, 423 (1970).

24. H. Harker, J. B. Horsley, and D. Robson, *J. Nucl. Mater.*, 37, 331 (1970).

25. J. Nwankwo and A. Turk, *Carbon*, 13, 495 (1975).

26. *JANAF Thermochemical Tables*, 2d. ed., NSRDS-NBS 37 (1971).

27. P. Magne and X. Duval, *Bull. Soc. Chim. Fr.*, A5, 1593 (1971).

28. H. Marsh and R. R. Adair, *Carbon*, 13, 327 (1975).

29. H. Abramowitz, R. Insinga and Y. K. Rao, *Carbon*, 14, 84 (1976).

30. G. W. Lee, *Coke and Gas*, 23, 398, 442 (1961).

31. J. W. Patrick and F. H. Shaw, *Fuel*, 51, 69 (1972).

32. A. Tomita, O. P. Mahajan, and P. L. Walker, Jr., preprints of *Fuel Chem. Div.*, ACS 173rd Mtg., 22(1), 4 (1977).

33. A. E. Cover, W. C. Schreiner, and G. T. Skaperdas, *Chem. Eng. Prog.*, 69(3), 31 (1973).

34. A. L. Kohl, R. B. Harty, and J. G. Johanson, *Chem. Eng. Prog.*, 74(8), 73 (1978).

35. C. A. Kumar, L. D. Fraley, and S. E. Handman, preprints of *Fuel Chem. Div.*, ACS 170th Mtg., 20(4), 260 (1975).

36. H. Amariglio and X. Duval, *Carbon*, 4, 323 (1966).

37. R. G. Jenkins, S. P. Nandi, and P. L. Walker, Jr., *Fuel*, 52, 288 (1973).

38. J. A. Cairns, C. W. Keep, H. E. Bishop, and S. Terry, *J. Catal.*, 46, 120 (1977).

39. D. W. McKee, *Carbon*, 17, 419 (1979).

40. H. Marsh and D. W. Taylor, *Abstracts 13th Biennial Carbon Conf.*, 1977, p. 46.

41. F. S. Feates, P. S. Harris, and R. T. K. Baker, *Proc. 7th Int. Congr. Electron Microscopy*, Grenoble, France, 1970, p. 357.

42. E. T. Turkdogan and J. V. Vinters, *Carbon*, 10, 97 (1972).

43. P. S. Harris, F. S. Feates, and B. G. Reuben, *Carbon*, 12, 189 (1974).

44. M. Letort and G. Martin, *Bull. Soc. Chim. Fr.*, 14, 400 (1947).

45. G. R. Hennig, *J. Inorg. Nucl. Chem.*, 24, 1129 (1962).

46. J. R. Fryer, *Nature*, 220, 1121 (1968).

47. R. T. K. Baker, R. D. Sherwood, and J. A. Dumesic, *Abstracts 14th Biennial Carbon Conf.*, 1979, p. 159.

48. Y. F. Chu and E. Ruckenstein, *Surface Sci.*, 67, 517 (1977).

49. S. Wong, M. Flytzani-Stephanopoulos, M. Chen, T. Hutchinson, and L. D. Schmidt, *J. Vac. Sci. Technol,* 14, 452 (1977).

50. D. W. McKee, R. Savage, and G. Gunnoe, Jr., *Wear,* 22, 193 (1972).

51. D. W. McKee and V. J. Mimeault, in *Chemistry and Physics of Carbon,* Vol. 8 (P. L. Walker, Jr., and P. A. Thrower, Eds.), Marcel Dekker, New York, 1973, p. 151.

52. J. W. Patrick, Coke Research Report No. 47, British Carbonization Research Association, Nov. 1967.

53. R. T. K. Baker and P. S. Harris, *Carbon,* 11, 25 (1973).

54. T. Buch, J. A. Guala, and A. Caneiro, *Carbon,* 16, 377 (1978).

55. P. S. Harris, F. S. Feates, and B. G. Reuben, *Carbon,* 11, 565 (1973).

56. D. W. McKee, unpublished observations.

57. C. Roscoe, *Carbon,* 6, 365 (1968).

58. J. M. Thomas and C. Roscoe, in *Chemistry and Physics of Carbon,* Vol. 3 (P. L. Walker, Jr., Ed.), Marcel Dekker, New York, 1968, p. 1.

59. J. M. Thomas and C. Roscoe, *Proc. 2nd Conf. Industrial Carbon and Graphite,* London, S.C.I., 1965, p. 249.

60. D. J. Allardice and P. L. Walker, Jr., *Carbon,* 8, 375 (1970).

61. D. J. Allardice and P. L. Walker, Jr., *Carbon,* 8, 773 (1970).

62. R. E. Woodley, *Carbon,* 6, 617 (1968).

63. R. E. Woodley, *Carbon,* 7, 609 (1969).

64. R. C. Asher and T. B. A. Kirstein, *J. Nucl. Mater.,* 25, 344 (1968).

65. K. Hedden, H. H. Kopper, and V. Schulze, *Z. Physik. Chem.* (Frankfurt), 22, 23 (1959).

66. B. P. Puri and R. C. Bansal, *Carbon,* 5, 189 (1967).

67. Yu. N. Vasil'ev, A. P. Martynov, and I. M. Bodrov, *Konstr. Mater. Osn. Grafita,* 8, 171 (1974); *Chem. Abs.,* 82, 101023h.

68. P. Hawtin and J. A. Gibson, *Proc. 3rd Conf. Industrial Carbon and Graphite,* S.C.I., London, 1970, p. 147.

69. J. R. Arthur and D. H. Bangham, *J. Chim. Phys.,* 47, 559 (1950).

70. E. Wicke, *Proc. 5th Int. Symp. Combustion,* Reinhold, New York, 1955, p. 245.

71. D. W. McKee, *Carbon,* 10, 491 (1972).

72. J. M. Thomas, in *Chemistry and Physics of Carbon,* Vol. 1 (P. L. Walker, Jr., Ed.), Marcel Dekker, New York, 1965, p. 1.

73. P. Magne, H. Amariglio, and X. Duval, *Bull. Soc. Chim. Fr.,* A6, 2005 (1971).

74. J. M. Thomas and P. L. Walker, Jr., *Carbon*, 2, 434 (1965).

75. G. W. Sears and J. B. Hudson, *J. Chem. Phys.*, 39, 2380 (1963).

76. J. M. Thomas and P. L. Walker, Jr., *J. Chem. Phys.*, 41, 587 (1964).

77. A. Szirmae, V. Rao, and R. M. Fisher, *Proc. 9th Int. Congr. Electron Microscopy*, 1, 452 (1978).

78. F. J. Long and K. W. Sykes, *J. Chim. Phys.*, 47, 361 (1950).

79. S. S. Barton, B. H. Harrison, and J. Dollimore, *J. Chem. Soc.*, *Faraday Trans. I*, 69, 1039 (1973).

80. C. Heuchamps, Thèse Ingénieur-Docteur, Université de Nancy, 1960.

81. E. A. Heintz and W. E. Parker, *Carbon*, 4, 473 (1966).

82. P. S. Harris, *Carbon*, 10, 643 (1972).

83. K. Motzfeldt, *J. Phys. Chem.*, 59, 139 (1955).

84. A. R. Ubbelohde and F. A. Lewis, in *Graphite and its Crystal Compounds*, Oxford, 1960, p. 143

85. D. Berger, B. Carton, A. Metrot, and A. Herold, in *Chemistry and Physics of Carbon*, Vol. 12 (P. L. Walker, Jr., and P. A. Thrower, Eds.), Marcel Dekker, New York, 1975, p. 1.

86. B. Neumann, C. Kröger, and E. Fingas, *Z. Anorg. Chem.*, 197, 321 (1931).

87. G. C. Bond, in *Catalysis by Metals*, Academic, New York, 1962, p. 447.

88. O. V. Krylov, in *Catalysis by Nonmetals*, Academic, New York, 1970, p. 176.

89. R. D. Weaver, M. Yasuda, A. E. Bayce, and L. Nanis, "Direct Electrochemical Generation of Electricity from Coal," *ERDA Report SAN-115-105-1*, May 1977.

90. K. Otto and M. Shelef, *6th Int. Congr. Catalysis*, London, 1976, paper B47.

91. F. S. Feates, P. S. Harris, and B. G. Reuben, *J. Chem. Soc.*, *Faraday Trans. I*, 70, 2011 (1974).

92. P. L. Walker, Jr., J. F. Rakszawski, and G. R. Imperial, *J. Phys. Chem.*, 63, 140 (1959).

93. B. P. Jalan and Y. K. Rao, *Carbon*, 16, 175 (1978).

94. Y. K. Rao, *Metall. Trans.*, 2, 1439 (1971).

95. I. Mochida, M. French, and H. Marsh, *Proc. 5th London Carbon and Graphite Conf.*, 1, 155 (1978).

96. E. Hippo and P. L. Walker, Jr., *Fuel*, 54, 245 (1975).

97. A. K. Burnham, *Fuel*, 58, 713 (1979).

98. D. W. McKee, *Fuel*, 59, 308 (1980).

99. J. M. Skowronski, *Fuel*, 56, 385 (1977).

100. J. Tashiro, I. Takakuwa, and S. Yokoyama, *Fuel*, 55, 250 (1976).

101. Y. Tamai, H. Watanabe, and A. Tomita, *Carbon*, 15, 103 (1977).

102. G. Blyholder and L. D. Neff, *J. Phys. Chem.*, 66, 1464 (1962).

103. A. J. Melmed, *J. Appl. Phys.*, 37, 275 (1966).

104. H. J. Grabke, *Carbon*, 10, 587 (1972).

105. M. K. Halpin and G. M. Jenkins, *Proc. Roy. Soc.* (London), A, 313, 421 (1969).

106. C. Wagner, unpublished, quoted in Ref. 93.

107. P. K. Lorenz and G. J. Janz, *J. Electrochem. Soc.*, 118, 1550 (1971).

108. J. L. Johnson, *Catal. Rev.-Sci. Eng.*, 14(1), 131 (1976).

109. S. P. Chauhan, H. F. Feldman, E. P. Stambaugh, and J. H. Oxley, preprints of *Fuel Chem. Div.*, ACS 170th Mtg., 20(4), 207 (1975).

110. S. P. Chauhan, H. F. Feldman, E. P. Stambaugh, J. H. Oxley, K. Woodcock, and F. Witmer, preprints of *Fuel Chem. Div.*, ACS 173rd Mtg., 22(1), 38 (1977).

111. C. A. Trilling, preprints of *Fuel Chem. Div.*, ACS 173rd Mtg., 22(1), 185 (1977).

112. A. Linares, O. P. Mahajan, and P. L. Walker, Jr., preprints of *Fuel Chem. Div.*, ACS 173rd Mtg., 22(1), 1 (1977).

113. K. Otto and M. Shelef, *Proc. 13th Biennial Carbon Conf.*, Irvine, Calif., 1977, paper CR-2(6), p. 50; *Fuel*, 58, 85 (1979).

114. E. J. Hippo, Ph.D. Thesis, Pennsylvania State University, 1977.

115. K. Otto, L. Bartosiewicz, and M. Shelef, *Fuel*, 58, 565 (1979).

116. K. A. Wilks, M.S. Thesis, Case Western Reserve University, 1974.

117. J. L. Johnson, Preprints of *Fuel Chem. Div.*, ACS 170th Mtg., 20(4), 85 (1975).

118. K. A. Wilks, N. C. Gardner, and J. C. Angus, preprints of *Fuel Chem. Div.*, ACS 170th Mtg., 20(3), 52 (1975).

119. R. Hahn and K. J. Huttinger, *Proc. 13th Biennial Carbon Conf.*, Irvine, Calif., 1977, paper CR-2(3).

120. P. P. Feistel, K. H. van Heek, H. Juntgen, and A. H. Pulsifer, *Carbon*, 14, 363 (1976).

121. N. Kayembe and A. H. Pulsifer, *Fuel*, 55, 211 (1976).

122. W. P. Haynes, S. J. Gasior, and A. J. Forney, in *ACS Advances in Chemistry Series* (L. G. Massey, Ed.), 131, 179 (1974).

123. W. G. Willson, L. J. Sealock, Jr., F. C. Hoodmaker, R. W. Hoffman, D. L. Stinson, and J. L. Cox, in *ACS Advances in Chemistry Series* (L. G. Massey, Ed.), 131, 203 (1974).

124. A. P. Malinauskas, *Chem. Eng. Progress Symp. Ser.*, 66, 81, 104 (1970).

125. M. R. Everett, D. V. Kinsey, and E. Romberg, in *Chemistry and Physics of Carbon*, Vol. 3 (P. L. Walker, Jr., Ed.), Marcel Dekker, New York, 1968, p. 289.

126. D. W. McKee and D. Chatterji, *Carbon*, 16, 53 (1978).

127. M. J. Veraa and A. T. Bell, *Fuel*, 57, 194 (1978).

128. H. Watanabe, M.S. Thesis, Tohaku University, Japan, A. Tomita (Trans.), Pennsylvania State University, 1974.

129. K. Otto and M. Shelef, *Carbon*, 15, 317 (1977).

130. A. Tomita, N. Sato, and Y. Tamai, *Carbon*, 12, 143 (1974).

131. R. T. Rewick, P. R. Wentrcek, and H. Wise, *Fuel*, 53, 274 (1974).

132. R. Hahn, K. J. Hüttinger, and P. Schleicher, *Proc. 5th London Int. Congr. Carbon and Graphite*, 1, 151 (1978).

133. D. J. Dwyer and G. W. Simmons, *Surface Sci.*, 64, 617 (1977).

134. S. D. Robertson, *Carbon*, 8, 365 (1970).

135. A. G. Oblad, *Catal. Rev.-Sci. Eng.*, 14(1), 83 (1976).

136. D. K. Mukherjee and P. B. Chowdhury, *Fuel*, 55, 4 (1976).

137. A. Tomita, O. P. Mahajan, and P. L. Walker, Jr., *Fuel*, 56, 137 (1977).

138. N. Gardner, E. Samuels, and K. Wilks, in *ACS Advances in Chemistry Series* (L. G. Massey, Ed.), 131, 217 (1974).

139. J. Weber and M. Bastick, *Comp. Rend. Acad. Sci.*, C, 280, 1177 (1975).

140. H. Marsh and D. W. Taylor, *Fuel*, 54, 219 (1975).

141. D. Brennan and P. C. Fletcher, *Trans. Faraday Soc.*, 56, 1662 (1960).

142. M. A. Qayyum and D. A. Reeve, *Carbon*, 14, 199 (1976).

143. A. Tomita and Y. Tamai, *J. Catal.*, 27, 293 (1972).

144. M. Boudart, A. W. Aldag, and M. A. Vannice, *J. Catal.*, 18, 46 (1970).

145. J. Sinfelt, *Catal. Rev.*, 3, 175 (1969).

146. J. Weber and M. Bastick, *2nd Int. Carbon Conf.*, Baden-Baden FRD, 1976, paper 39.

147. J. Weber and M. Bastick, *High Temp. High Press.*, 9(2), 177 (1977).

148. S. D. Robertson, N. Mulder, and R. Prins, *Carbon*, 13, 348 (1975).

149. E. Keren and A. Soffer, *J. Catal.*, 50, 43 (1977).

150. A. J. Robell, E. V. Ballou, and M. Boudart, *J. Phys. Chem.*,
      68, 2748 (1964).

151. G. Hennig, *J. Chim. Phys.*, 58, 12 (1961).

152. J. Abrahamson, *Carbon*, 11, 337 (1973).

ELECTRONIC TRANSPORT PROPERTIES OF GRAPHITE, CARBONS, AND RELATED MATERIALS

Ian L. Spain[*]

Laboratory for High Pressure Science and Engineering Materials Program
Department of Chemical and Nuclear Engineering
University of Maryland
College Park, Maryland

[*]Current affiliation: Department of Physics, Colorado State University, Fort Collins, Colorado.

I. INTRODUCTION

The electrical resistivity, ρ, of materials spans an enormous range
of values, which is greater than that for any other material property.
Some representative data are summarized in Figure 1. Materials based
on carbon have resistivity across this range. Even pure carbon can
occur in a highly insulating, crystalline form, diamond, and a nearly
metallic (semimetallic) form, graphite. It is speculated that a
metallic form of carbon is stable at very high pressure ($\gtrsim 200$ GPa)
[A21,A24]. When graphite is intercalated with strong acids, such as
AsF_5, SbF_5, room temperature conductivity as high as that of copper
has been found [Q81]. Intercalated compounds of graphite were aptly
called *synthetic metals* by Ubbelohde [Q67,Q51]. At least one of
these metals (C_8K) is a superconductor at low temperature [Q39,Q47].

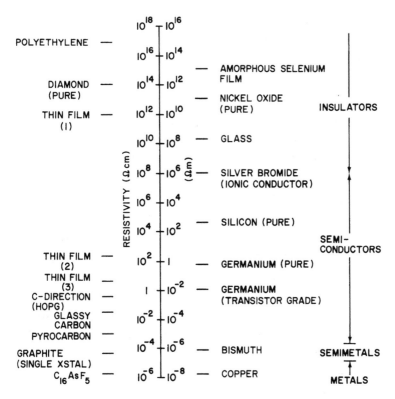

Figure 1. Comparison of the room temperature resistivities of
various forms of carbon with other materials.

The electrical conductivity, σ, can be related to two important
properties of the material--the density of current carriers, N, and
their mean mobility, μ (see Appendix A).

$$\sigma = N e \mu \equiv \rho^{-1} \tag{1}$$

In crystalline materials it is usually differences in N which control
the enormous range of values of σ or ρ. For instance, $N \sim 10^{25}\ m^{-3}$
in graphite at room temperature, but $\lesssim 10^{7}\ m^{-3}$ in naturally occurring
diamond at room temperature. However, their mobilities differ only
by a factor of ~ 10. Further changes in conductivity occur for a

given material as defects are introduced. Some representative data
for the basal resistivity as a function of temperature for several
types of carbon and graphite are illustrated in Figure 2.

Carbon is unique in that the degree of disorder can be modified,
and in some cases controlled, over wide ranges. On the one hand
there are just about perfect crystals of graphite, while on the other
there are nearly amorphous carbons. Between these extremes materials
can be prepared with various degrees of long- and short-range order.
Also, the bonding between carbon atoms can differ.

The anisotropy of the crystal structure of graphite also introduces important phenomena. For instance, in polycrystalline graphite,
the crystallites can be aligned as in highly oriented pyrolytic
graphite, or more-or-less randomly oriented as in some nuclear-grade
materials. Since heat treatment is an important step in the preparation of most carbons, thermal stresses caused by the anisotropic
thermal-expansion coefficient are important. Carbons generally are
nonequilibrium materials with high residual stresses, as is typically
the case with other anisotropic ceramics.

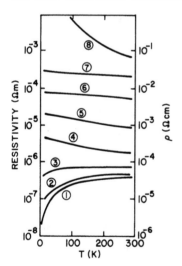

Figure 2. Resistivity vs. temperature plots for various forms of
carbon: (1) single crystal; (2) highly oriented pyrolytic graphite;
(3) graphite whisker; (4) pyrolytic carbon; (5) petroleum coke carbon;
(6) lampblack base carbon; (7) glassy carbon; (8) carbon film--
electron beam evaporated.

Another unusual feature of the electronic properties of carbons stems from the fact that graphite is a *semimetal*. Accordingly, the modifications to the electronic structure that occur as a result of structural modifications are significantly different from those in an insulator, semiconductor, or metal.

Apart from its scientific interest, the electrical conductivity of carbon-based materials is of great importance because of its industrial applications. Brushes for electrical motors are constructed of polycrystalline graphite because of a combination of electrical and lubricating characteristics. High-temperature applications include furnace cores and electrodes for metallurgical applications. Carbon resistors are used as electrical components, and automobile ignition cables are often constructed with carbon cores to maximize suppression of radio frequency noise. At the present time, synthetic metals are being investigated for applications for which high conductivity, low density, and high strength are desirable. A good review of industrial applications can be found in Ref. A4.

Another reason for interest in the electrical-transport properties is that they give a characteristic "fingerprint" by which the material can be classified. In many cases a measurement of the electrical resistivity, along with its temperature coefficient, can be used as a guide to the development of the structure of the carbon. This method was used, for instance, in the development of highly oriented pyrolytic graphite (HOPG) [F6,F8].

The present review attempts to survey the electrical-transport properties of these carbon-based materials, without attempting to be an exhaustive study. It is aimed at chemists or engineers who wish to gain a working knowledge of these materials, along with some insight into the reasons for particular properties or trends occurring. To maximize continuity, a number of important concepts are treated separately, in appendices. Furthermore, the reference list at the end of the chapter is organized by topic and therefore functions also as a bibliographic guide to more specific information.

For physicists, this survey may broaden their knowledge of electrical-transport properties to a wider class of materials;

however, they will not find a deep discussion of present concerns of active researchers, such as the mechanism of localization in amorphous carbons, or the possibility of excitonic effects in crystalline graphite.

The relationship between structural and electrical characteristics is the underlying concern of this discussion. First, a brief description is given of the main features of the electronic structures of diamond, graphite, and carbon. Then a major class of materials—heat-treated carbons—is more fully ennumerated, since many important features of the transport properties can be introduced in regard to this class, and overall concepts seen clearly. Third, a discussion follows of specific materials and properties.

The International System of Units is used. For convenience, alternative units for the quantities most often encountered are compared in Table 1. Whenever convenient, graphs include both S.I. and alternative, commonly used units.

II.   ELECTRONIC STRUCTURE AND BONDING OF GRAPHITE, DIAMOND, AND CARBONS

It is well known that carbon atoms can form chemical bonds by rearranging the configuration of the outer electrons. This rearrangement is called *hybridization* and affects the $2s^2$ and $2p^2$ electrons. One possible hybridization scheme is found in diamond where four tetrahedral bonds are formed, each at an angle of 109° 28' to each other (Figure 3a). These very strong, covalent bonds consist of hybridization of one s state and three p states (sp^3 bonding). The diamond structure (Figure 3b) is one of several possible structures that satisfies this bonding arrangement. The overall symmetry is cubic.

By analogy with molecules, the possible electronic states of the sp^3 electrons in diamond can be divided into bonding and antibonding states, the latter having higher total energies. In diamond at $T = 0K$, all $4N$ bonding electrons (N = number of atoms in the crystal) lie in the lowest possible energy states, which are

Table 1. Units in the International System, Compared with Other Unit Systems

Quantity		S.I.	Other Units	
Conductivity	σ	$1\ \Omega^{-1}\ m^{-1}$	$= 0.01\ \Omega^{-1}\ cm^{-1}$	$= 9 \times 10^9$ e.s.u.
Resistivity	ρ	$1\ \Omega m$	$= 100\ \Omega cm$	$= 1.11 \times 10^{-10}$ e.s.u.
Mobility	μ	$1\ m^2\ V^{-1}\ sec^{-1}$	$= 10^4\ cm^2\ V^{-1}\ sec^{-1}$	$= 3 \times 10^6$ e.s.u.
Carrier density	N	$1\ m^{-3}$	$= 10^{-6}\ cm^{-3}$	
Magnetic field density	B	1 Tesla	$= 10^4$ gauss	
Hall coefficient	R_H	$m^3\ C^{-1}$	$= 10^6\ cm^3\ C^{-1}$	$= 10^{13}/9$ e.s.u.

(a)   (b)

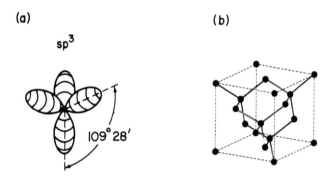

Figure 3. (a) Illustration of the electron charge cloud density for
a carbon atom with sp^3 hybridization. (b) Crystal structure of
diamond, showing the cubic cell.

broadened into a band of levels (Figure 4). The concept of density
of states, D(E), is discussed in Appendix B, but it can be defined
as the number of energy states allowed to the electrons in unit volume
of crystal (m^3) in unit energy interval, (eV).

The bonding states are separated from the antibonding states by
approximately 5.3 ev, so that pure diamond is a very good insulator.
For diamond to be able to conduct electricity, electrons must be
excited across this band gap into the conduction band (antibonding
band). This requires very high temperatures to achieve even a small
density of electrons and holes for conduction. However, small concentrations of impurities or defects can increase the density of one
type of carrier (electron or hole), so that the conductivity of

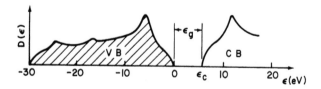

Figure 4. Hypothetical sketch of the density of states of diamond.
Zero energy is taken at the top of the valence or bonding band (VB).
The bottom of the conduction or antibonding band (CB) is at 5.3 eV.
(Figure interpolated from band calculation of Hemstreet, Fong, and
Cohen [C11].)

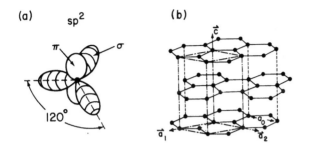

Figure 5. (a) Sketch illustrating the electron charge density for sp^2 hybridization. (b) The crystal structure of graphite, showing the unit cell.

natural or synthetic diamond is much lower than the theoretical value for pure diamond (see Section XIV).

Another possible hybridization scheme for the bonding electrons occurs if the carbon atoms arrange themselves in hexagonal rings. This is the case with graphite and aromatic molecules. Orbitals composed of one s and two p orbitals (sp^2 bonds) are formed at 120° to each other in a plane. These form the strong covalent bonds between carbon atoms in the plane and are known as σ bonds. The one remaining orbital has a p_z configuration (Figure 5a), called the π orbital, which provides the weak bonding between adjacent layers in the graphite structure. This weak interplanar bonding is sometimes incorrectly referred to as Van der Waals bonding. The overlap of the π orbitals on adjacent atoms in a given plane provides the electron bond network responsible for the high mobility of graphite.

The electronic density of states for graphite is illustrated in Figure 6a. The σ orbitals form bonding and antibonding bands with a large band gap between them. However, the π bands just touch for two-dimensional graphite (a single layer of carbon atoms) but overlap by ~40 meV for three-dimensional[*] graphite (layer planes stacked with the ...ABAB... sequence illustrated in Figure 5b). It is this small overlap with makes graphite a semimetal. The density of states near

_____

[*]Hereafter, "2D" and "3D" will often be used for two-dimensional and three-dimensional, respectively.

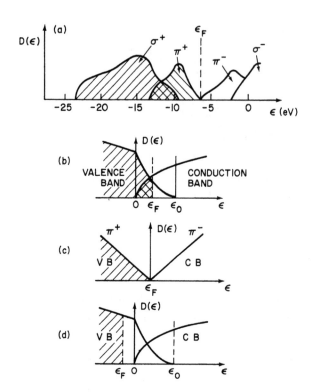

Figure 6. (a) Sketch of the density of states for graphite. Zero energy is taken at the vacuum level. (b) Density of states for 3D graphite near the region of band overlap. Zero energy is taken at the bottom of the conduction (π^-) band. (c) Density of states for a single sheet of carbon atoms (2D graphite) near the band touching region. (d) Density of states for 3D graphite doped with ~100 ppm boron.

the region of the overlap of the π-bonding band (valence band) and π-antibonding band (conduction band) is illustrated in Fig. 6b. There are $4N$ bonding electrons (N = number of atoms), $3N$ of which are in the σ^+ band. If the π bands just touched, as in 2D graphite (Figure 6c), all of the N remaining electrons would lie in the valence band at T = 0K, and the conduction band would be empty (Figure 6c). However, in 3D graphite a few electrons (~$10^{-4} N$) lie in the conduction band, leaving exactly the same number of empty states, called holes, in the valence band.

If there are impurities or defects in the crystal, the density
of holes, P, and conduction electrons, N, can be different (N ≠ P).
For instance, boron produces acceptor levels lying below the conduction-
band edge, so that electrons are removed from both the conduction and
valence bands to the acceptor levels (Figure 6d). The case drawn
corresponds to a doping density of boron of ~100 ppm at T = OK, where
the Fermi level (see Appendix C) is shifted below the bottom of the
conduction band. Only holes can conduct electricity at this temperature. This will be considered in greater detail in Section XI.

Further changes in the electronic structure can occur as the
lattice becomes more defective. A discussion of this is left to
Appendix D and the following section, where the specific case of heat-
treated carbons is considered.

Another hybridization scheme for carbon pertains for long chain
molecules in which two electrons form bonds between neighboring carbon
atoms in the chain (sp-bonding), while two electrons remain free for
bonding side chain molecules and groups. A wide range of polymeric
materials can be formed. Perhaps the simplest is polyethylene. Although polymers are outside the scope of this discussion, it should be
mentioned that polyethylene is an excellent insulator, because of a
very large gap between bonding and antibonding states. Actually, this
energy gap can be modified by attaching the appropriate side group to
the molecule. Extensive research is being carried out today on the
electrical properties of materials of this kind. Two possible goals
of such research are the development of synthetic metals and of high-
temperature superconductors [A7 and A12].

III. PRELIMINARY DISCUSSION OF HEAT-TREATED CARBONS

A. Structural Features

A wide range of heat-treated carbons can be prepared by heating an
organic precursor to a temperature in excess of ~1000°C. The heat
treatment must be carried out in a nonoxidizing atmosphere (vacuum,
inert gas, or reducing atmosphere) to avoid combustion. In this

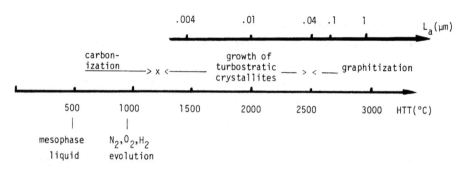

Figure 7. Sketch of the carbonization and graphitization process
for a graphitizable carbon.

section some of the major variables in the carbonization process will
be explored and important terminology introduced.

The heat-treatment temperature (HTT) is the temperature to which
the material is subjected. The carbonization and graphitization processes are kinetic in nature, so the length of time of application of
HTT is also important. Although changes may appear to be imperceptible
after a long time at a fixed temperature, further changes do occur but
with decreasing rate (approximately logarithmic). The heat-treatment
condition (temperature and time) can be expressed in terms of an
equivalent time at different temperatures [B4 and B13].

With increasing HTT at fixed residence time, organic precursors
tend to form chars in the temperature region ~500-900°C. N_2 and O_2,
for example, are evolved by ~1000°C and H_2 by ~1200°C. Sulfur is not
evolved until even higher temperatures, and certain metallic impurities (e.g., Fe) may require a leaching process to be eliminated. The
gradual elimination of impurities by heat treatment is called *carbonization*.

At sufficiently high HTT some carbons *graphitize*--that is, form
crystallites with a three-dimensional (...ABAB...) stacking sequence.
The development of the structure of a graphitizing carbon with increasing HTT is sketched in Figure 7. The organic precursor melts
and then solidifies for HTT ≲500-700°C. The liquid *mesophase* has an
important influence on the subsequent graphitization process, since

(a) (b)

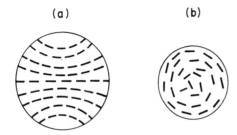

Figure 8. Sketch of the arrangement of aromatic networks in
(a) graphitizable mesophase spherule; (b) carbon-black spherule.

the planar aromatic molecules aggregate into spheres in which the
molecules align themselves nearly parallel to neighbors, but perpendicular to the surface of the sphere (Figure 8). Growth of spheres
can then occur readily by coalescence.

For HTT $\gtrsim 1200°C$, when the carbon is more or less pure, aromatic
ring networks grow and stack in relation to each other, forming
crystallites. A sketch showing the arrangement of these crystallites
is given in Figure 9, while Figure 10 diagrams the growth of the average

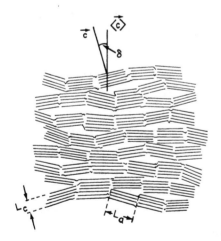

Figure 9. A hypothetical cross section through a turbostratic
graphite; < c > is the principal axis, and δ is the angle between
the c axis of a specific turbostratic crystallite and < c >. A
bulk, heat-treated carbon is usually isotropic, so the concept of
a principal axis is inapplicable. (Figure adapted from [A8]).

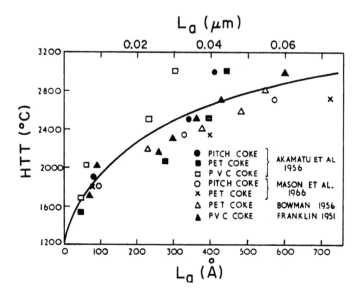

Figure 10. A sketch of the variation of turbostratic crystallite dimension L_a with HTT for various precursors. (Figure adapted from [A8].)

crystallite dimension, L_a, as a function of heat-treatment tempera-ture, for several precursors. Neighboring crystallites are tilted with respect to one another, and the sample taken as a whole is iso-tropic. Voids between crystallites ensure that the density is lower than that of perfect graphite (2.267 g cm^{-3}).

It is to be noted that an aromatic molecule with only 50 C atoms has a benzene-ring structure with a carbon-carbon bond length very close to that of graphite (1.42 Å). However, x-ray diffraction data show that the interplanar separation (d) is greater than that of single-crystal graphite (3.354 Å). Over a fairly wide range of HTT d ~3.44 Å, which is characteristic of well-developed layer planes that are stacked parallel to each other but which are rotated at random with respect to each other, about the c axis of the crystal-lite. This random rotational arrangement of planes was termed *turbo-stratic* by Biscoe and Warren [B2]. In this case, use of the term *crystallite* to define a small region of material with parallel planes (see Figure 8) is misleading. Because the term is so widely used, the practice is continued here. However, materials with turbostratic crystallites are not graphites. Rather, they are carbons.

If HTT exceeds a certain limit (e.g., 10 min treatment above 2800°C for anthracene-based carbons) the planes begin to stack in the regular ..ABAB.. sequence of single-crystal graphite. This process is *graphitization*. Turbostratic packing of the planes may be inferred from the presence of an assymetric (002) x-ray diffraction peak (see References A8, B5, and B7). As graphitization proceeds three-dimensional reflections (hkl) appear and the d spacing decreases toward its crystalline value. Several different measures of the degree of graphitization, γ, have been used [A8] but a particularly simple one is based on the change of d spacing [G12].

$$\gamma = \frac{d_{max} - d_{actual}}{d_{max} - d_{min}} \tag{2}$$

where

d_{max} = value for turbostratic carbon (~3.44 Å)

d_{min} = value for single-crystal graphite ($= c/2$, where c is the unit cell parameter, 6.708 Å)

d_{actual} = value measured for the sample

In single crystals and highly graphitized materials such as HOPG, regions between crystallites can be characterized in terms of dislocations [B1]. However, sp and sp^3 bonds, impurity atoms, dangling bonds, and voids become increasingly important [B15] for materials that are less well graphitized, or for carbons with lower HTT. It is to be noted that interstitial atoms and defects in the layer plane can increase the value of d above 3.44 Å.

Many types of organic precursors do not graphitize when heat treated. Even at 3000°C the layer size, L_a, may not exceed 50 Å. All thermosetting polymers fall into this category, forming disordered carbons with glasslike properties. Normally these materials

are porous because of the large quantities of gas evolved, much of
it trapped in the pores. Processes have been developed to control
the porosity, allowing a material to be made that has many useful
applications [A10]. This class of carbons is referred to as nongraphitizing, and is sometimes characterized as glassy, or vitreous.

The relative inability of some carbons to graphitize is believed
to be due, at least at lower heat-treatment temperatures, to the
presence of strong sp^3 crosslinking bonds between entangled graphitic
ribbons (see Section VIII). The sp^3 bonds can open an energy gap
between bonding and antibonding states, although the disordered nature
of the material also produces a mobility gap discussed in Appendix D.
At higher HTT the inability of the planes to grow and stack is due to
the tangled arrangement of the planes.

Most graphitizing carbons are mechanically soft and are sometimes
referred to as *soft carbons*. Nongraphitizing carbons are mechanically
hard, and are sometimes referred to as *hard carbons*. In this review,
the terms "graphitizing" and "nongraphitizing" carbons will be used
exclusively.

Another type of carbon is carbon black [A24,R4]. Carbon blacks
are formed by the condensation of carbon from the gas phase into small
spheres containing between $\sim 10^5$-10^9 atoms (Figure 8) corresponding to
sphere diameters $\sim 10^{-2}$-10 μm. However, the aromatic layers are arranged
parallel to the surface. Accordingly, it is difficult for spheres to
coalesce, and carbon blacks only partially graphitize.

Carbons heat treated to a condition where they are relatively
pure are often referred to as *cokes*. The percentage of volatile impurities may be as high as 15% for cokes exposed to only 500°C, known
as a "green" or "raw" coke. Often a calcining process is used to
preshrink (densify) the coke if it is to be used in the manufacture
of polycrystalline graphite (see Section VII). Another term encountered is "needle-coke." This type can be prepared when the precursor
pitch has a low concentration of insoluble particles, so that only a
few mesophase spheres are nucleated. Extensive regions are formed
with an ordered arrangement of planes. A needle coke initially breaks

up into needles when mechanically ground, then into very small
platelets (~100 μm diameter) like other coke.

Nature provides an abundant supply of impure carbons in the form
of coal [A23]. The carbonization process of the precursor material
(rotted vegetation) has taken place over a very long period of time
under conditions in which impurities cannot easily escape. A wide
variety of coals can be found, and further information is available
elsewhere [A23].

B. Evolution of Electronic Properties of Heat-
   Treated Carbons

In the previous subsection the evolution of the structure of heat-
treated carbons was briefly discussed. Parallel to the structural
development are changes in the electronic transport properties, which
are summarized for a specific graphitizable carbon in Figure 11
(T = 300 K). Major features of interest are:

1. In the region of low HTT ($\lesssim$900°C for 15 min residence time)
    the resistivity drops dramatically, then levels off. The
    magnetoresistance is extremely small (see Appendix B for a
    discussion of galvanomagnetic effects).

2. · In the HTT range between ~900–2100°C the resistivity remains
    roughly constant, then begins to fall to its single-crystal
    value, 4 x 10^{-7} Ωm.

3. The Hall coefficient changes sign at HTT $\gtrsim$1500°C, becoming
    positive at higher temperature, reaching a peak value at
    ~2000°C before falling, changing sign again, and tending
    to its single-crystal value ~-0.5 μCm^{-3}.

4. The magnetoresistance (magnetic field is arbitrarily taken
    as 0.3 T) is negative over a range of HTT, then becomes
    positive in the HTT range where graphitization occurs.

The earliest model for the change of electronic and structural
properties of graphitizable carbons was put forward by Mrozowski
[H13]. A later modification of this model is indicated in Figure 11.
At a heat-treatment temperature of ~600°C the π-energy levels of the
aromatic hydrocarbons were assumed to coalesce into a broad valence

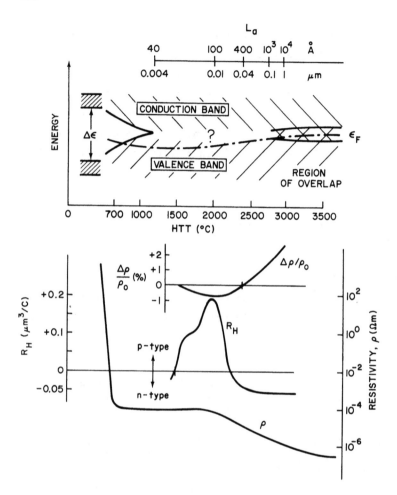

Figure 11.   Summary of Mrozowski's model for the evolution of the
electronic properties of heat-treated carbons.   Qualitative data on
the "crystallite dimension" L_a, the Hall coefficient R_H, the magnetoresistance at 0.3 T, $\Delta\rho/\rho_0$, and the resistivity ρ are included.

and conduction band with a band gap of ~0.3 eV between them.   As a
function of increased heat-treatment temperature the energy gap decreases until at ~1300°C it closes.   Mrozowski assumed that a two-
band model (see Appendix B) is applicable for both graphitic materials
and carbons.   For graphitizing material treated above 2200°C a small

band overlap appears, developing to the value of ~40 meV for single-crystal graphite [A18].

However, it was shown that the properties of soft carbons heat-treated between about 1300–2500°C could not be explained in terms of a simple two-band model [H1]. This is the region indicated by a question mark by Mrozowski in Figure 11, characterized by negative magnetoresistance. It is possible that carriers are increasingly localized by defects and crystalline disorder as HTT decreases.

The theoretical problem of the change of electronic structure as a graphite changes to a defective turbostratic carbon is unsolved. McClure and Ruvald [C16] showed that the energy bands near the Fermi level of ideal turbostratic graphite were essentially the same as those of a single sheet of graphite (2D graphite). This is particularly important for the electronic states near the Fermi energy. In 2D graphite the band overlap disappears, and the density-of-states curve is modified (see Figure A-15 and Appendix C). However, in the region of partial graphitization no models have been developed to estimate whether the band overlap closes to a zero value for ideal turbostratic carbon. The material is inhomogeneous on a microscopic level at least. Thermodynamically, the Fermi level is required to be constant throughout the material so that charge transfer must take place between regions of different graphitization. The boundary regions could be most important in controlling the electrical conductivity, in both these and more disordered materials.

Furthermore, the layers are not ideally turbostratic. They have finite size (L_a) and contain many defects, such as vacancies, interstitial clusters, dislocations, or grain boundaries; also they are in a stressed, nonequilibrium state. It may be surmised that the energy bands develop tails of localized states (see Appendix D, Figure A-19), which become more important as the crystallite size decreases, or the defect concentration increases. However, there are no theoretical models to justify these suppositions.

In the following, the transport properties of several types of carbons and graphites will be considered in greater detail. Not all

of these materials fall into the category of heat-treated carbons,
but it will be seen that materials prepared in different ways can
behave similarly. The discussion will begin with nearly perfect
materials, then proceed to more defective.

IV.   SINGLE-CRYSTAL AND HIGHLY ORIENTED PYROLYTIC GRAPHITE

Nearly perfect crystals of graphite can be found in nature, natural
crystals (NCG), or precipitated from iron-carbon solutions, kish
graphite (KG). Some samples of pyrolytic graphite (PG) heat-treated
to very high temperatures under restraining stresses also consist of
well-oriented 3D crystallites. Although such synthetic materials may
be prepared in different ways, they will be referred to collectively
as highly oriented pyrolytic graphite (HOPG) (see Ref. F5 for a
review).

The electronic structure of NCG and HOPG have been probed near
the Fermi energy using a number of sensitive techniques (see Ref. A18
for a review). The dispersion relationship $\epsilon(k_x, k_y, k_z)$ of the carriers (see Appendix C) is the same for these materials, even for
subtle details, such as the value of the band overlap parameter Δ
which depends sensitively on the long-range stacking order of the
planes [C30]. However, the mechanisms by which carriers are scattered in HOPG and KG or NCG are different, particularly at low temperature where intrinsic (electron-phonon) scattering mechanisms are
reduced. Consequently, certain properties of these materials, such
as the Hall coefficient at low magnetic field, are dissimilar. Although the galvanomagnetic properties of graphite are not fully understood at the present time a reasonable picture for the influence of
defects and impurities has been developed and it is with this aspect
that we will be mainly concerned.

Natural crystals of graphite (NCG) can be found in a number of
locations, such as Madagascar, the USSR, and the Ticonderoga area of
the U.S. In the latter case, fairly large crystals (~several mm in

the basal planes, ~1 mm in the c direction) can be obtained. A
typical flake may contain several twinning planes, but a sample for
transport measurements can usually be cut from a single-crystal
region (NSCG). Great care must be taken during the cutting and lead-
attachment procedure, since these crystals are very fragile. It is
recommended that a single crystal be examined before and after cutting
to ensure that damage has not occurred (for discussion of cutting
procedures for HOPG see Ref. F9).

Naturally occurring crystals of graphite contain appreciable
concentrations of impurities, which can be partially removed by chemical leaching. Typical treatments are prolonged boiling in concen-
trated hydrofluoric acid or heating (~2000°C) in flowing fluorine gas.
The two treatments may be carried out successively. After such treatment, the crystals contain ppm levels of metallic impurities such as
Fe.

Single crystals of graphite have been prepared by recrystalliza-
tion from Fe-C melts and are known as kish graphite (KG). Impurities,
principally Fe, must be removed as for NSCG. The largest crystals of
KG are usually of the order of ~1 mm platelets, but may be larger in
exceptional cases (see Ref. E2).

HOPG is a material whose properties form a natural bridge between
single-crystal material and pyrolytic graphites and carbons. It consists of many small crystallites whose c axes are well aligned but
whose a axes are randomly oriented. The crystallites form a mosaic
pattern, and crystallite boundaries can be described in terms of
arrays of dislocations [B1].

HOPG is used extensively for measurements of physical properties
and as the starting material for intercalation compounds of graphite
(see Section XV). The material normally used is formed by hot-pressing
pyrolytic graphite at elevated temperature. C axes may be aligned to
within ~0.2° of the average c axis, and crystallites are typically
$\gtrsim 1$ μm in the basal plane (L_a) and ~0.1 μm in the c direction (L_c).
This material is used as a monochromator for x rays or thermal neutrons [F5]. Annealing at ~3500°C under an applied c-axis stress
results in further crystallite growth (L_a ~4-10 μm).

Electrons in perfect crystals are described by extended wave functions over the whole crystal. Without defects the conductivity would be infinite. Resistance is provided by intrinsic and other defects that scatter the electrons. At finite temperature in an otherwise perfect crystal, the displacement of atoms from their equilibrium positions--thermal vibrational displacements--provides an intrinsic resistance, which vanishes at T = OK even though there is a zero-point vibrational motion of the atoms. Extrinsic defects that provide an additional source of resistance include impurities, point defects, dislocations, and grain-boundaries (see Figure A-2 in Appendix A).

In the case of a polycrystalline material such as highly oriented pyrolytic graphite, the electronic wave functions extend over the crystallite, but the wave functions in one crystallite are not coherent with respect to neighboring crystallites. Using Heisenberg's uncertainty principle, the finite extent of the crystallite, L, imposes an uncertainty on the wave vector, $k = 2 \pi/\lambda$ where λ is the electron wavelength, given by

$$\Delta k \ L^{-1} \sim 1 \tag{3}$$

This does not impose serious limitations on the use of band structure concepts until L is reduced to hundreds of interatomic distances. Accordingly, the boundaries between crystallites are considered to act as scattering sites for the electrons and contribute to the resistivity. It is stressed that 3D ordering must remain in the crystallites for these remarks to be applicable.

The electrons and holes which contribute to conductivity are those with energy close to the Fermi energy, ε_f, ($\sim \varepsilon_f \pm 3$ kT). These carriers lie close to the corners of the Brillouin zone (Appendix C and Ref. C20) in highly elongated constant energy surfaces. In intrinsic graphite the densities of holes (P) and electrons (N) are equal, but impurities such as boron and defects such as those introduced by neutron irradiation, can create unequal densities (see Appendix B.1). The carrier density is very small in graphite, only

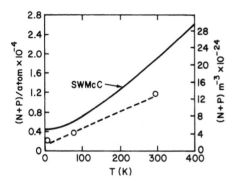

Figure 12. The variation of the density of carriers calculated from the SWMcC model for the carrier properties [F2]. Values calculated from the formula $(N + P) \equiv \sigma/e\mu_M$ for HOPG are also indicated [F2]. The experimental curve for turbostratic carbon is also indicated, by the dotted line [G12].

about one carrier for 10^4 atoms (Figure 12). This compares with one free carrier per atom in copper. This is again a consequence of the semimetallic nature of graphite.

The effective masses of holes and electrons are small in the basal plane, varying markedly along the zone edge, and lying in the range $m_{\perp}^* \sim 0.06 - 0.004\ m_0$.[+] In the c direction they are much higher, lying above $5\ m_0$. Accordingly, the conductivity anisotropy is expected to be at least 100, even for an isotropic relaxation time (see Equations 1 and 9 in Appendix A). Also, the small basal effective masses are consistent with very high mobilities for electron currents in the basal plane, since

$$\mu = \frac{e\tau}{m_{\perp}^*} \tag{4}$$

$$\left[\text{units} \equiv \frac{Csec}{kg} \equiv m^2\ Vsec^{-1} \right]$$

where τ is the relaxation time for scattering (see Appendix A). At room temperature the mean mobility of carriers in graphite is $\sim 1\ m^2\ V^{-1}\ sec^{-1}$, compared to $\sim 4 \times 10^{-3}\ m^2\ V^{-1}\ sec^{-1}$ in copper.

[+]A discussion of the effective mass concept is given in Appendix C.

The dispersion relationship for the carriers developed by
Slonczewski-Weiss and McClure (SWMcC) is complicated, since the
carrier properties vary markedly along the zone edge and the constant
energy surfaces are trigonally warped in the k_x-k_y plane (Figure A-14).
Calculations of the conductivity and galvanomagnetic effects have
usually used simple models for the relaxation time and dispersion
relationship (see Ref. A18). Accordingly, our theoretical understanding is still not completely quantitative. Some current topics
that are controversial are discussed in Refs. D3 and D4.

Measurements of the basal resistivity of several samples of NCG
have been carried out by Soule [E15], and are reproduced in Figure 13.
The resistivity decreases with temperature. Near room temperature,
$\rho(T)$ is approximately linear

$$[(1/\rho \frac{d\rho}{dT}) \sim (12 \times 10^{-4} K^{-1})]$$

A knee is apparent in the curve near 150 K, then at lower temperature
($\lesssim 10$ K) the curve flattens. Curves of $\rho(T)$ for natural crystals and
kish graphite are similar, and differences between samples are only

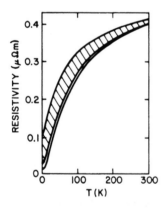

Figure 13. Curve of the resistivity as a function of temperature for
single-crystal graphite (sample EP-14 [E15]). For comparison the
shaded area is representative of HOPG with resistivity ratios between
~4 (upper bound) and 18 (lower bound). (Data from [F2] and [F8].)

Table 2.   Comparative Data for Natural Single Crystals[a], Kish
           Graphite[b], and High Quality HOPG[c]

Quantity	Temp.	Units	Sample			
			EP-14	EP-7	HOPG[d]	KG[e]
ρ	298	$\Omega m \times 10^{-7}$	4.1	4.1	3.95	~6
	77.4		2.0	2.0	2.205	—
	4.2		0.106	0.235	0.215	~.133
μ	298	$m^2 \, V^{-1} \, sec^{-1}$	1.54[f]	1.50	1.173	~1.2
	77.4		9.4	8.9	6.68	—
	4.2		115	83	126.3	~100
(N + P)	298	$m^{-3} \times 10^{24}$	9.9 (14.5)[g]	10.1 (14.5)	13.5	—
	77		3.3 (4.3)	3.5 (4.7)	4.25	—
	4.2		5.2 (5.7)	3.2 (4.2)	2.30	—
$\rho_{298}/\rho_{4.2}$			37.9	17.1	18.3	45

[a]Source: From Ref. E15.
[b]Source: From Ref. E10.
[c]Source: From Ref. F2.
[d]Sample 1, [F2] $\overset{*}{M}$(Equation 7).
[e]Source: From Ref. E10.
[f]Values for μ for EP-14 and EP-7 are derived from Soule [E15],
  page 706.
[g]Values in parentheses for EP-7 and EP-14 refer to values given
  by McClure [D11].

apparent at low temperature, where carrier scattering is dominated
by collisions with impurities.

    The resistivity ratio $\rho(298)/\rho(4.2\ K)$ can be used as a rough
guide to sample purity, similar to metals [A25]; but for graphite,
the resistivity ratio is smaller than the mobility ratio $\mu(4.2\ K)/$
$\mu(298\ K)$ because the density of carriers varies with temperature
(Figure 12). Some representative data for resistivity ratios are
given in Table 2, values as high as 45 having been reported, corresponding to mobility ratios ~200 [E10].

Samples of HOPG with the largest basal crystallite dimension
have resistivity, and mobility curves very like those of NSCG or KG.
The highest resistivity ratio found for HOPG is $\rho(298)/\rho(4.2\ K)$ =
18.3, corresponding to a mobility ratio $\mu(4.2\ K)/\mu(298\ K)$ = 107 [F2].
The basal crystallite dimension L_a for this specimen was ~10μm, based
on the assumption that grain boundary scattering is the dominant
factor controlling mobility at 4.2 K (see Fig. A-2, Appendix A).

The influence of crystallite size on the resistivity can be best
illustrated through the variation of mobility with temperature for
samples of different perfection. Curves of the magnetoresistance
mobility for several specimens are plotted in Figure 14. Intrinsic
(electron-phonon) scattering leads to a mobility variation $\mu\ \alpha T^{-1.2}$
over the temperature range ~50-300 K, and the curves are somewhat

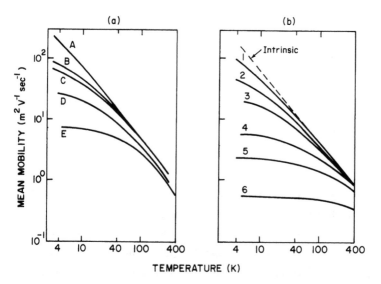

Figure 14. (a) Experimental data for the variation of magnetoresistance mobility $\mu_M(.3T)$ with temperature. (A) NSCG, sample
EP-11[E15]; (B) NSCG, sample EP-14 [E15]; (C) high quality HOPG
[F2,F8] and NSCG, sample EP-7 [E15]; (D) HOPG with smaller crystallite size [F8]; (E) pyrolytic carbon [G12]. (b) Calculated curves
for $\mu_M(T)$ for HOPG compared to perfect graphite: 1) λ_d = 4 μm,
2) λ_d = 1 μm, 3) λ_d = 0.4 μm, 4) λ_d = 0.1 μm, 5) λ_d = 0.04 μm,
6) λ_d = 0.1 μm.

steeper for the temperature range ~20-40 K [ZE1]. However, at the lowest temperatures $\mu(T)$ flattens due to scattering by defects (ionized impurities for NCG and KG, grain boundaries and dislocations for HOPG).

The influences of intrinsic and defect scattering can be separated using Mathiesson's rule (see Appendix A.2).

$$\frac{1}{\mu_{total}} = \frac{1}{\mu_i} + \frac{1}{\mu_d} \tag{5}$$

where μ_i is the intrinsic and μ_d the defect mobility. For scattering by grain boundaries, μ_d is approximately independent of temperature, and curves for $\mu_{total}(T)$ for different values of the mean free path for scattering by defects, λ_d, (see Appendix A.2) are constructed on this basis in Figure 14. It is the crystallite dimension L_a that effectively controls λ_d in HOPG, since the carrier velocity component in the c direction is very small compared with that in the basal plane. For approximately cylindrical crystallites, L_a and λ_d are related by $L_a \sim 2.5\ \lambda_d$, where L_a is the diameter of the crystallite (see Appendix A.2).

From the curves in Figure 14 it is to be noted that the room temperature value of μ is not appreciably altered until $\lambda_d \lesssim 0.4\ \mu m$. Accordingly, the room temperature resistivity is not increased much beyond its intrinsic value $\rho_{298\ K} = 4.0 \times 10^{-7}\ \Omega m$ until the resistivity ratio $\rho_{298\ K}/\rho_{4.2\ K}$ falls to ~2.

Since the above materials are characterized by near equality of carrier densities (N-P) and mobilities ($\mu_n \sim \mu_p$), the simple two-band formula,

$$\left(\frac{\Delta\rho}{\rho_0 B^2}\right)^{1/2} = \mu_M \tag{6}$$

has been used extensively to estimate carrier mobility. Unfortunately, the value of μ_M depends on the value of magnetic field at which

Equation 6 is used, so that there is an arbitrariness in the value
of $\overline{\mu}_M$. Reasons for this are discussed in Appendix B.5. The most
useful formula for $\overline{\mu}_M$ is

$$\mu_M^* = \left(\frac{\Delta\rho}{\rho_0 B^2}\right)^{1/2}_{B^*} \tag{7}$$

where B^* is the magnetic field at which $\Delta\rho/\rho_0 = 1$, i.e. the resistance doubles. It should be pointed out that μ_M^* is roughly two times
higher than the conductivity mobility defined by

$$\mu_c = \frac{\sigma}{(N + P)e} \tag{8}$$

and there is no satisfactory explanation for this. A detailed discussion is given in [D26] and [D3]. The discrepancy can be seen in
Figure 12 where values of (N + P) evaluated from $(N + P)_M \equiv \sigma/e\mu_M$
are compared with values calculated from the band model.

Differences in the scattering mechanisms and carrier density
differences (P - N) between different types of graphite can result
in profound differences in the Hall effect. Curves of $R_H(B)$ of Soule
are compared in Figure 15 with data for a high quality sample of HOPG.
Data for KG have recently been published by Tsuzuku and co-workers
[E9]. The variation of $R_H(B)$ was considered quantitatively by Soule
[E15] and McClure [D11] in terms of nearly compensated bands of electrons and holes, with small pockets of mobile minority carriers.
However, it can be seen that changes in the low-field Hall constant
are negative-going for NSCG at 4.2 K, but positive-going at 77 K.
For HOPG, rather broader negative-going trends are evident at all
temperatures. Accordingly, it was not possible to relate the pockets
of minority carriers to the band model. An alternative explanation
has recently been given in terms of trigonal warping of the constant
energy surfaces, warping which produces low field behavior similar
to that from small pockets of mobile minority carriers [D4]. Differences in behavior of NCG and HOPG were then explained in terms of
different scattering mechanisms for the carriers.

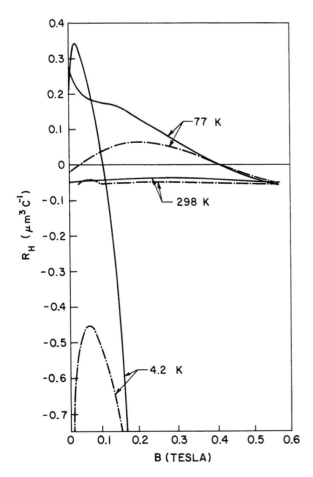

Figure 15. Curves of the Hall coefficient, R_H, vs. magnetic field for NSCG (full line) and HOPG (dot-dash line). (Data from [E15] and [F8].)

In the temperature region below ~60 K the overall curves of $R_H(B)$ are shifted to negative values compared to NCG or KG. Furthermore, at all temperatures the curves assume more negative values for samples of HOPG with smaller crystallite dimension L_a. This is consistent with the scattering of carriers from the crystallite boundaries, which is more effective for hole-type carriers than electron because the holes have higher average speeds in the basal plane [D25]. Curves of mobility ratios $b = \mu_n/\mu_p$ for NCG and HOPG are compared in

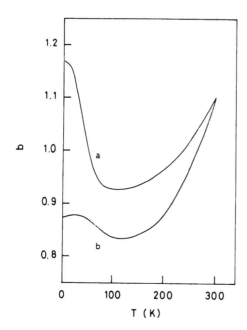

Figure 16. Variation of the mobility ratio b = μ_n/μ_p with temperature for (a) HOPG [F8]; (b) natural single crystal of graphite [E15].

Figure 16, showing this trend. By comparison, hole mobility is
higher than electron for NCG or KG at low temperature, since scattering by charged impurities is more effective for the slower carriers
(electrons). The curve of b(T) for NCG in Figure 16 is compatible
with this hypothesis.

At room temperature R_H is n-type and similar for NSCG, KG and
HOPG. Apparently the electrons become increasingly more mobile than
the holes as temperature increases above ~200 K. This is an intrinsic
phenomenon, controlled by the band properties and electron-phonon
interactions.

In Appendix B it is pointed out that at high field R_H should
saturate, giving an estimate of carrier compensation (P - N). Unfortunately, galvanomagnetic data cannot be fitted to the STBM model
(see Appendix B), or reasonable extensions of this model [F2]. However, (P - N) can be estimated from high field data, particularly the
conductivity component ($\sigma_{xy}B$) (see Ref. N7). The results indicate

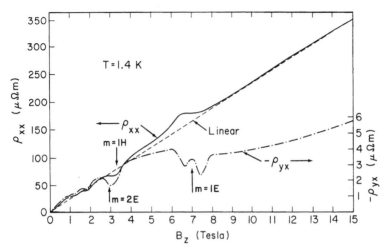

Figure 17. Experimental data for the variation of the normal resistivity, ρ_{xx}, and the Hall resistivity ($\rho_{yx} = R_H B$) with magnetic field for a high quality sample of HOPG [F2].

that in this one respect HOPG is more "perfect" than NCG, since (P - N) is smaller for HOPG. Higher values for NCG are presumably related to impurities. The acceptors in HOPG are possibly related to structural defects, but this has not been established (see Section XII for discussion of the effect of neutron irradiation on the Hall effect). Boron-doping produces positive values of R_H (see Section XI).

At low temperature (T $\lesssim$ 20 K) oscillations appear in curves of $\rho(B)$ or $R_H(B)$ (Figure 17). These are the Shubnikov-de Haas oscillations which are associated with the condensation of electronic states into magnetic energy levels (see Appendix C). The Shubnikov-de Haas effect has been of particular value for the evaluation of the shape and size of the Fermi surface [A18]. The amplitude of oscillations depends on the defects and impurities in the sample and is closely related to the mobility values for holes and electrons, confirming the discussion given above for the differences in mobility ratio for NCG and HOPG.

Graphite is very unusual since all the carriers associated with the transport properties lie in the ground-state magnetic energy

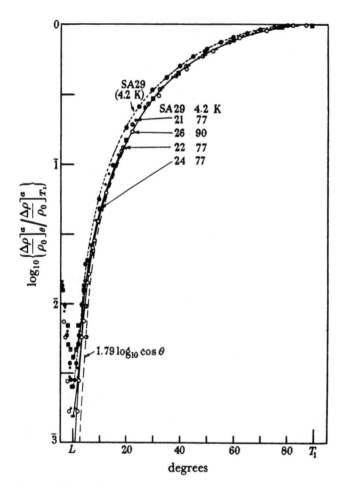

Figure 18. Plot of the logarithm of the relative magnetoresistance as a function of the angle between the magnetic field and the electric current vector in the basal plane. Data for several specimens are included [F8].

level for magnetic fields above ~7 T. This region is called the *quantum limit* [T4]. Exploratory studies in this region can be found in References, Section T.

The magnetoresistance anisotropy of these materials is consistent with a very high degree of preferred orientation for the c axes. A plot of a magnetoresistance rotation diagram for HOPG is given in Figure 18. The magnitude of the anisotropy ratio $\Delta\rho/\rho_0$(B11C)/

$\Delta\rho/\rho_0$($B\perp C$) depends weakly on the applied magnetic field and temperature, because of the complexity of the factors controlling magnetoresistance. Typical values lie in the range ~100-600 for B ~0.5 T
[F8]. At 4.2 K, for a sample of HOPG with average misorientation
$< \delta > $ ~0.5°, a magnetoresistance anisotropy ratio of 170 is expected
for B = 0.3 T. Magnetoresistance anisotropy of KG and polycrystalline
samples of natural graphite are discussed in Refs. D14, D15, E6.

V. PYROLYTIC GRAPHITES AND CARBONS

Pyrocarbons are prepared by pyrolyzing a hydrocarbon at elevated
temperature. Deposition temperatures are practically limited to the
range between ~1800-2500°C. Further heat treatment can be carried
to close to the sublimation temperature (~3600°C), and those heat
treated to $\gtrsim$3000°C are nearly completly graphitized. The deposition
temperature (T_d) can be a misleading parameter, since pyrocarbons are
often deposited onto a heated substrate whose temperature increases
as the deposit thickens in order to maintain constant surface temperature. In some cases, for example, the inner layers of the deposit can
reach temperatures as high as 3600°C while the surface is at 2200°C.
These inner layers may be in a highly stressed state due to differential thermal expansion that helps to improve alignment. The resulting
material is silvery and has properties close to those of natural single
crystals. It was this observation that led Ubbelohde and co-workers
to hot-press pyrolytic deposits [F6].

Other factors of importance in the deposition process are the
pressure and flow rate of gas, which can affect the formation of soot
nuclei (see Ref. G1). Dopants, such as silane, can be introduced into
the gas. These impurities affect the deposition and graphitization
processes [M17,M18].

Klein has presented a synopsis of his laboratory's work on the
properties of these materials in Volume 2 of this series [G15]. Here,
a brief summary of the properties will be given.

The main types of structural disorder in these materials can be
roughly summarized as follows:

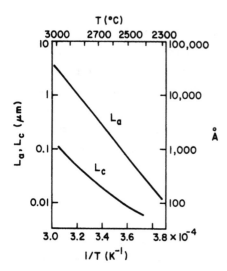

Figure 19. Variation of the turbostratic crystallite dimensions of pyrolytic carbons deposited at 2200°C. Inner layers reached higher temperatures, allowing crystallites to grow. Values of L_a were deduced from x-ray, electrical resistivity, and thermal conductivity measurements. L_c values are from x-ray diffraction. (Data adapted from [G21].)

1. Crystallite size L_a increases with deposition temperature or subsequent heat-treatment temperature. Data of Morant are summarized in Figure 19 for samples prepared by a resistance-heating technique so that inner layers of the deposit were at higher temperature [G21].

2. Each crystallite is tilted with respect to the next. The material appears to be wrinkled. C axes of material deposited at ~2200°C have an angular spread of ~20°. The decrease in the angle of misalignment with increased temperature is sensitive to the strain condition of the material. An x-ray determination of the preferred orientation for the same materials, as in Figure 19, is illustrated in Figure 20. Preferred orientation sets pyrocarbons apart from heat-treated bulk carbons.

3. The spacing between adjacent planes is approximately 3.42 Å for T_d between 1700–2300°C. This is characteristic of a turbostratic arrangement of planes. The d spacing decreases with increasing heat-treatment temperature, becoming very close to that of single-crystal graphite (3.354 Å) for heat-treatment temperature of 3500°C. Using the index of graphitization, γ, in Equation 2, graphitization is relatively complete at this temperature.

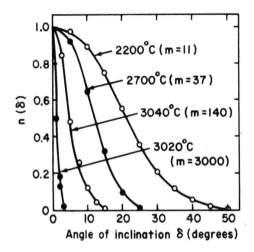

Figure 20. The orientation parameter n(δ) for the same pyrolytic carbons as in Figure 19. Curves are fitted to $\cos^m\delta$, with m-values indicated. Note that temperature is not the only parameter of importance, but also the state of stress. (Data adapted from [G20].)

4. Closely related to the decrease of d spacing is the increase of crystallite size L_c. In material heat treated to 3500°C, $L_c > 0.1$ μm (Figure 20).

5. Within crystallites there are defects such as vacancies, interstitials, dislocations, impurities. At lower deposition temperature, particularly, the inclusion of hydrocarbons and soot nuclei becomes more important.

Some workers refer to all materials deposited above ~2000°C as graphites. This is incorrect. The correct term is *pyrocarbon*, until graphitization (formation of 3D crystallites) is relatively complete (≳3000°C for typical residence times).

The room temperature electrical resistivity as a function of deposition temperature is sketched in Figure 21. A large drop in ρ(298 K) occurs for deposition temperatures in the range ~1800-2000°C. Although the crystallite dimension varies in a smooth logarithmic fashion (see Figure 19), it is important to note that room-temperature mobility is not strongly reduced until $\lambda_d \lesssim 0.1$ m (Figure 14). Thus, the increase in ρ(300 K) occurs for the temperature range $T_d \lesssim 2200$°C corresponding to reduction of L_a below ~0.25 μm (Figure 19).

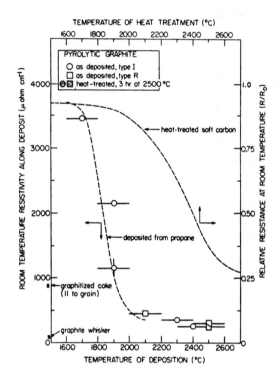

Figure 21. Variation of the room temperature basal resistivity of pyrocarbons as a function of deposition temperature compared to soft carbons as a function of heat-treatment temperature (Figure adapted from [G1].)

Contributory factors which increase ρ are the increased turbostratic disorder and crystallite misalignments.

Typical data of the variation of resistivity with temperature for pyrocarbons are given in Figure 22. Quite clearly pyrolytic graphite heat treated to the highest temperature has characteristics of single-crystal graphite. As the room temperature resistivity increases, the positive temperature coefficient decreases, becoming negative for deposition temperatures below ~2500°C. The temperature dependence is more clearly shown in Figure 23 and compared with other materials.

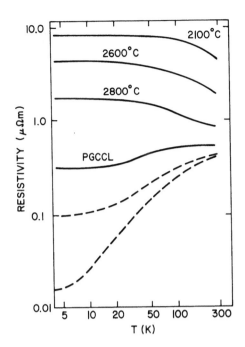

Figure 22. Variation of resistivity with temperature for a series of pyrocarbons deposited at 2100°C and subsequently heat treated. The sample marked PGCCL is a hot-pressed graphite with relatively small crystallite size (L_a ~0.4 μm), while the dashed curves are for PG heat treated to ~3500°C [G12]. (Curve adapted from [G7] with data from [G12].)

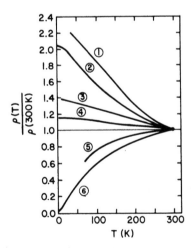

Figure 23. The temperature dependence of the resistivity for various types of carbons including pyrolytic: (1) pyrocarbon (2500°C) (2) pyrocarbon 2400°C; (3) pyrocarbon 1900°C; (4) coke and lamp-black base carbons (HTT = 2000°C); (5) annealed pyrocarbon 3000°C (6) annealed pyrographite (3500°C), single-crystal graphite and high quality HOPG. (Figure adapted from [G12].)

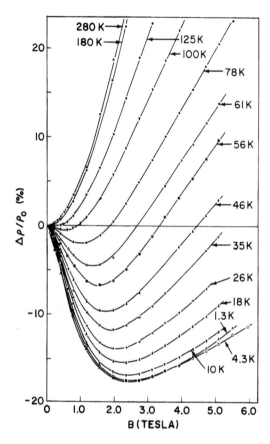

Figure 24.   The magnetoresistance at several temperatures for a
sample of pyrocarbon deposited at 2100°C and heat treated at 2300°C.
(Figure adapted from [G7].)

        The interpretation of the galvanomagnetic effects is complicated
by the appearance of negative magnetoresistance (see Appendix B.4).
The most extensive measurements on pyrocarbons have been made by
Delhaes, Kepper, and Uhlrich [G7] for materials deposited between
2100°C and 2800°C.   Figure 24 illustrates the change in character of
the magnetoresistance for a sample heat treated at 2300°C, as the
measurement temperature varies between 280 and 1.3 K, indicating that
the negative component generally strengthens as temperature is lowered.   At room temperature the magnetoresistance is positive in all
magnetic fields for this specimen.

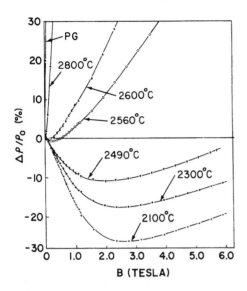

Figure 25. The variation of magnetoresistance at 4.3 K for several pyrocarbons deposited at 2100°C and heat treated to temperatures indicated. (Figure adapted from [G7].)

Magnetoresistance curves at 4.3 K for several specimens are given in Figure 25. Negative magnetoresistance increases as deposition temperature decreases. By analogy with heat-treated soft carbons the negative magnetoresistance is expected to diminish for further reduction of deposition temperature (see Section VI).

At room temperature the positive magnetoresistance for several samples is given in Figure 26. The magnetoresistance mobility in $m^2 \, V^{-1} \, sec^{-1}$ may be directly obtained by taking the square root of the magnetoresistance at B = 1T (Equation 6) as accomplished on the right of the figure. Note that μ_M drops by a factor of ~40 from the single crystal value (1.2 $m^2 \, V^{-1} \, sec^{-1}$) to the sample deposited at 1900°C ($\mu_M \sim 0.03 \, m^2 \, V^{-1} \, sec^{-1}$), while the resistivity increases by about the same factor (10 $\mu\Omega m$ compared to 0.4 $\mu\Omega m$). The increase in resistivity as measurement temperature is lowered for HTT $\lesssim$2000°C in Figure 21 is associated, then, with the decrease in carrier density with temperature (Figure 12), while the increase in mobility with decreasing measurement temperature (see Figure 14, $\lambda_d \lesssim 0.1$ m) is

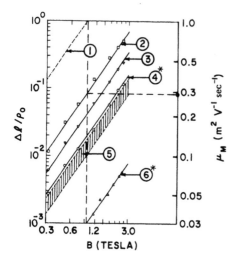

Figure 26. Logarithmic plot of the magnetoresistance vs. temperature
for several pyrolytic carbons compared to single crystal and graphitized coke. Magnetoresistance mobilities are estimated at the right
of the figure by following a horizontal line from the B = 1T value,
as accomplished for sample (2) for instance. Samples with an asterisk
were inductively heated and have much higher average crystallite mis-
orientations. (Figure adapted from [G12].)

insufficient to produce a positive temperature coefficient of resistivity. Accordingly, for the samples in Figure 22, negative temperature coefficient of resistivity is *not* to be associated with a band
gap or mobility gap.

The Hall coefficient at room temperature is sketched in Figure
27 as a function of deposition temperature compared with heat-treated
soft carbons (to be discussed in the next section). For a specimen
deposited at 1900°C, the Hall curve is near the maximum, with Hall
mobility, $\mu_H \sim 0.02$ m^2 V^{-1} sec^{-1}, which is comparable to the magnetoresistance mobility (~ 0.03 m^2 V^{-1} sec^{-1}). The two-band model is not
strictly applicable in this case (see Appendix B) but mobility values
are expected to be roughly correct (e.g. within a factor of ~ 2-3).
Thus, the total density of carriers is roughly the same for 1900°C
deposited material as for single-crystal material. In the pyrocarbon
the carriers are predominantly holes and the Fermi energy is lowered
due to acceptor defects.

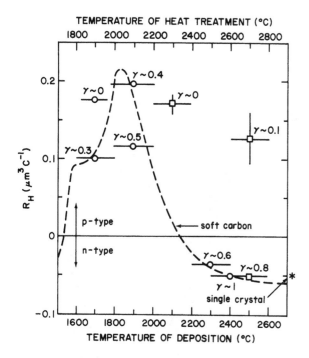

Figure 27. The variation of Hall coefficient for pyrolytic carbons as a function of deposition temperature. The dashed line represents results for soft carbons with heat-treatment temperature at the top of the figure. O = as-deposited, inductance heated; □ = as-deposited, resistance heated. [Figure adapted from [G12].)

The Hall curve (Figure 27) shows clearly that deposition tempera-ture is not the only factor that controls the properties of pyrolytic carbons. The samples, as deposited, with resistance heating behave differently from those inductively heated and retain lower values of the degree of graphitization, γ. Inductively-heated samples follow closely the curve for heat-treated soft carbons, and show an increase in γ. Note that the deposition temperature of graphitizable pyrolytic deposits corresponds to a heat-treatment temperature for carbons that is about 200°C higher. A difference of 200°C is not significant since residence times may differ and the deposition temperature may be higher than stated as a result of thermal gradients.

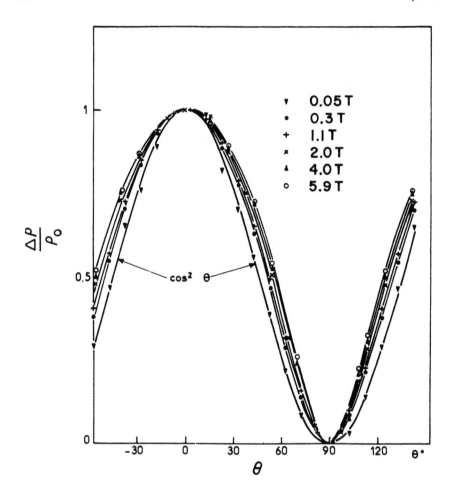

Figure 28. Magnetoresistance rotation diagram for pyrolytic carbon
(HTT = 2800°C) at 4.3 K compared with a $\cos^2\theta$ term. The secondary
transverse effect is very small. (Figure adapted from [G7].)

The mean crystallite misorientation angle may be estimated from
magnetoresistance anisotropy measurements (Appendix B) in the region
of positive magnetoresistance. A typical plot is illustrated in
Figure 28. It is to be noted that the curves vary as the measurement field is changed, as a result of the complicated magnetic field

dependence of the magnetoresistance. Over a limited range of magnetic
field values, $\Delta\rho/\rho_0$ can usually be fitted to a power law B^n, and the
index n depends on temperature (n ~1.8 at 300 K, see Figure 26, falling
to ~1.0 for some samples at 4 K, see Figure 17). The reason for this
behavior is not understood, but the magnetic field dependence of $\Delta\rho/\rho_0$
can be correlated well with the dependence on angle, θ.

Pyrocarbons in the region of negative magnetoresistance can
exhibit complicated magnetoresistance anisotropy diagrams, $\Delta\rho/\rho_0(\theta)$,
as the normal component of magnetic field, B cos θ = B_{11}, changes from
a value where negative magnetoresistance prevails, to another regime
where positive magnetoresistance prevails, as in Figure 24 for T = 56 K.
Examples of this behavior are given in Ref. G7.

VI. GRAPHITIZABLE CARBONS

The evolution of the structure of heat-treated, graphitizable carbons
has already been discussed in Section III, and the changes of electronic properties outlined. In this section a more detailed discussion will be given.

Graphitizable carbons heat treated above ~1800°C have properties
that roughly parallel those of pyrocarbons. One major difference is
the alignment of crystallites in pyrocarbons, discussed in Section V.
Figure 21 compares the variation of resistivity with treatment or
deposition temperature for these two types of carbon. The relatively
sharp increase in resistivity for HTT $\lesssim$2600°C is again attributable
to the reduction in crystallite dimension L_a. The relative displacement of the two curves for pyrocarbons and heat-treated carbons is
related to the preferred orientation of the crystallites in PC. It
is also noted that the deposition temperature for PC may be inaccurate,
due to temperature gradients, and the residence time longer. The
plateau in $\rho_{300\ K}$(HTT) for HTT $\lesssim$1900°C is not fully understood, since
this is the region where disorder controls the electronic properties.

The variation of the Hall coefficient and magnetoresistance as a
function of HTT (15 min residence time) for a series of soft carbons

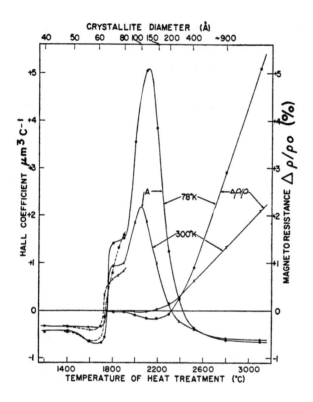

Figure 29. The variation of the transverse magnetoresistance and
Hall coefficient at 300 and 77 K for a soft carbon as a function of
heat-treatment temperature. (Figure adapted from [H18].)

is illustrated in Figure 29. Mrozowski [H13] proposed that the Fermi
level is depressed below the intrinsic value by acceptor-type defects
so that holes dominate the conduction processes. This explains the
positive values of the Hall coefficient.

Electron spin resonance experiments show that the density of
states at the Fermi energy, $D(\varepsilon_F)$, reaches a minimum value at HTT
~2600°C and maximizes at ~1200°C [H16], (Figure 30). The shift of
ε_F with HTT in Figure 11 is consistent with these latter data. ESR
experiments also show that the density of localized spin centers
roughly parallels the increase in $D(\varepsilon_F)$ (Figure 30). Electrons are

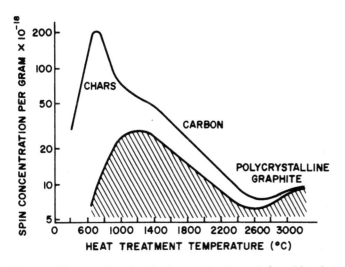

Figure 30. A sketch of the variation of localized carrier concentration (upper curve) and the density of free carriers (hatched curve) for heat-treated carbons, as obtained from electron spin resonance data. (Figure adapted from [H17].)

probably trapped at vacancies, although this has not been established theoretically. This is consistent with results on irradiated carbons and graphite, discussed in Section XII.

Changes in the magnetoresistance roughly parallel those in pyrocarbons. Data at low temperature ($2 \lesssim T \lesssim 30$ K) are illustrated in Figure 31 for carbons heat treated to different temperatures. The magnetoresistance is negative in this temperature range for HTT $\lesssim 2400°C$ and reaches a minimum value for the samples shown in Figure 31 at HTT = 2200°C. Note that the negative component of magnetoresistance diminishes below about 3 K for some specimens in Figure 31, similar to that observed for pyrocarbons (Figure 24).

Although the development of negative magnetoresistance appears to be closely related to the turbostratic arrangement of layer planes (see Appendix B.4), Hishiyama [H6] showed that negative magnetoresistance was not associated with localized spins, as suggested by Toyozawa [A20], since neutron-irradiated samples of soft carbon were shown to have more pronounced negative magnetoresistance after annealing. As

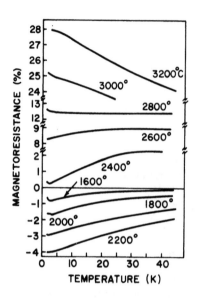

Figure 31. Temperature dependence of the magnetoresistance of a
soft carbon treated at different temperatures. Data are recorded
at B = 2.2T. (Curve adapted from [H9].)

discussed in Section XII, annealing removes localized carriers introduced by irradiation.

In the early stages of heat treatment ($\lesssim$1000°C for 15 min residence time used by Mrozowski and co-workers [H18]) the material can
be considered to consist of many organic molecules with different
molecular weights. The concept of a valence or conduction band is
not applicable in this case. However, an energy gap exists between
bonding and antibonding states, which decreases as the fused-ring
molecules increase their molecular weight until at ~1200°C the energy
gap closes. For HTT in the range ~600–1200 the energy states of the
molecules may be considered to fuse into bands of levels, and by HTT
~600°C the gap would be ~0.3 eV.

The dramatic change in the character of the conductivity can be
seen in Figure 32. The conductivity decreases sharply for HTT $\lesssim$700°C
and the curves for 300 K and 4.2 K diverge, consistent with a large
increase in carrier density or mobility at higher heat-treatment
temperature. Similarly the Hall coefficient falls in the HTT range

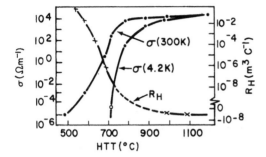

Figure 32. Variation of conductivity and Hall coefficient for an anthracene char as a function of heat-treatment temperature. The Hall coefficient changes from positive for HTT $\lesssim 700°C$ to negative for HTT $\lesssim 900°C$. (Figure adapted from [H2].)

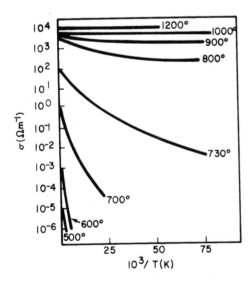

Figure 33. Arrhenius plot for the variation of conductivity with temperature for the same samples as Figure 32. (Figure adapted from [H2].)

where σ changes, indicating a reduction in the carrier density at
lower HTT.

Carmona et al. [H2] developed the idea that these noncrystalline
carbons are undergoing an insulator-to-metal transition. The reduction in the band gap can be seen clearly from the curves of ln σ vs.
1/T (Figure 33). In a small range of temperature ($\sim$1-10 $\overset{\circ}{K}$) the conductivity of materials heat treated at 900-1000°C appears to be characterized by Mott-type variable range hopping (see Appendix D for a
discussion). Similar behavior will be discussed for nongraphitizable
carbons in Section VIII.

VII. BONDED POLYCRYSTALLINE CARBONS AND GRAPHITES

A material commonly referred to as *polycrystalline graphite* is the
structural material used for furnace components, nuclear reactors,
brushes, etc., which is made by mixing a petroleum coke flour with a
coal-tar pitch binder. This mixture is molded or extruded to the
desired shape ("green" state), then baked at a temperature typically
$\sim$1000°C. Finally it is heat-treated in the range $\sim$2500-3000°C (see
Ref. A9 for an example). This material is partially turbostratic
although certain samples prepared from needle cokes are highly
graphitic. The crystallites may be partially aligned, although the
anisotropy of the electrical resistance is typically less than $\sim$2.
The anisotropy depends critically on the extrusion process.

The preparation conditions, structure, and properties of these
materials depend on their application. In many instances mechanical
properties at elevated temperature are most important, since these
materials have relatively high strengths above 2000°C compared to
metallic alloys, which are normally much stronger at lower temperature. Apart from the process conditions outlined above, other steps
may be included, such as calcining the coke, impregnating before
graphitizing to decrease porosity, hot-working to reduce pore and
crack size (which improves mechanical properties, etc. Nuclear
materials must undergo a purification step. Many processes are
proprietary.

Mrozowski studied the electrical properties of bonded carbons in the green and baked condition [I16]. He found that the resistance was largely controlled by the bridges of binder material between the coke particles, since the resistivity of the binder was ~30 times higher. A relationship between the resistivity of the bonded material and the proportions of filler and binder coke was proposed

$$\rho = \frac{B}{d_f^x \, d_b^{1/2}} \tag{9}$$

where d_f is the density of filler (mass of all filler particles in a unit volume of the bonded material) and d_b the density of the binder (mass of binder particles in unit volume of the bonded material). B is a factor proportional to the real resistivity $\overline{\rho}$ of the filler coke particles, and x an exponent. This relationship was found to give a satisfactory account of the variation of resistivity of these materials with x ~ 2.0 [I16]. Other semiempirical relationships between the bulk resistivity and the reciprocal of the mean layer plane diameter have been found.

The high temperature resistivities of several graphitic materials heat treated to 3000°C are shown in Figure 34 and compared with a pyrolytic carbon heat treated to the same temperature. For the latter material the resistivity coefficient is positive throughout the temperature range up to 2500°C, while bonded graphites have higher resistivities and a change of temperature coefficient of resistivity at ~1000 K. The curves can be roughly explained as follows.

1. The density of carriers increases throughout the temperature range, and above ~100 K is roughly linear with T (see Figure 12). If the mobility only varies weakly with temperature, a drop in resistivity occurs.

2. The mobility does not vary strongly with temperature until the mean free path for intrinsic scattering processes (electron-phonon) becomes comparable to that from defect scattering. This occurs at a relatively low temperature for pyrocarbon heat treated to 3000°C (see Figure 14, λ_d ~ .4 μm), but for a bonded carbon λ_d may be as small as .01 μm, so that intrinsic processes only become important

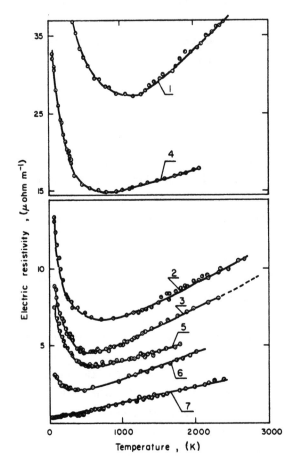

Figure 34. The variation of resistivity with temperature for a series of polycrystalline graphites and carbons: 1) low density (1.0 gm cm^{-3}), 2) density = 1.9 gm/cm^3 (parallel), 3) density = 1.9 gm/cm^3 (perpendicular), 4) density = 2.0 gm/cm^3 (parallel) (heat-treated under pressure at 2400°C), 5) density = 2.0 gm/cm^3 (perpendicular) (heat-treated under pressure), 6) density = 2.1 gm/cm^3 (heat-treated at 2400°C with metal catalyst), 7) pyrolytic carbon (deposition temperature 2100°C). All samples were subsequently heat treated to 3000°C, and direction specified is relative to overall c axis. (Figure from [I10].)

near ~1000 K. Above this temperature the decrease in mobility outweighs the increase in carrier density so resistivity
increases with measurement temperature.

3. The temperature at which resistivity reaches a minimum value
    should shift upwards with overall resistivity, in agreement
    with the experimental curves, since λ_d decreases as ρ in-
    creases.

4. Differences in the resistivity above ~2000°C are related to
    porosity and crystallite misorientation. Increases in either
    of these variables increases resistivity. Results were ob-
    tained in Figure 34 for current flow parallel and perpendicu-
    lar to the average c axis of the bulk sample.

The curves shown in Figure 34 are similar to others reported in
the literature for materials prepared under somewhat different conditions (see references in Section I of the Bibliography). Further
data on boron-doped polycrystalline carbons and graphites are presented in Section XI.

VIII. NONGRAPHITIZABLE CARBONS

As discussed in Section III.A certain organic precursors, such as
thermosetting polymers, do not graphitize. A sketch of a possible
structural model is given in Figure 35 for PVDC carbon heat-treated
to 2700°C. Although the networks are shown in the form of ribbons,
they are probably more isometric--that is, extend roughly equal distances in orthogonal directions within the carbon network. However,
although the carbon network may be extensive, for material heat
treated to ~3000°C there are relatively flat, unstrained portions,
such as P in Figure 35, that extend only ~50 Å. Planes can be bent
or twisted. In some respects the structure resembles that of high-
strength, hard carbon fibers (Section IX.). This analogy has been
stressed by Jenkins et al. [J7]. The main difference between the
two structures is the higher degree of preferred orientation found
in fibers, allowing layer-plane growth to proceed more readily.
There is controversy about the existence of tetrahedral, sp^3 bonds in
glassy carbon. A brief discussion of this topic is given by McKee
in a review [A10].

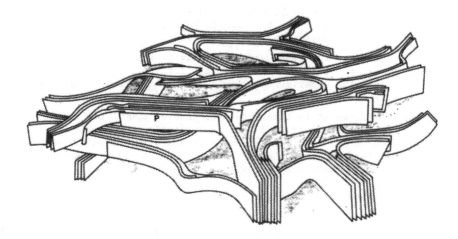

Figure 35. Proposed structure for a glassy carbon prepared from
polyvinylidene heat treated to 2700°C. (Figure kindly supplied by
Dr. H. Marsh from unpublished work.)

Typical properties of commercial glassy carbons include a density
of 1.3-1.55 g/cm^3, apparent porosity from 0-12%, and room temperature
electrical resistivity of 10-50 μΩm. The layer spacing within the
microfibrils is 3.4 Å, characteristic of turbostatic packing. The
fiber bundles must be more tightly packed than those in Figure 35.
X-ray analysis gives a crystallite dimension, L_a, of the order of
10-30 Å for material heat treated to 1100°C, and 30-50 Å for HTT =
2800°C.

Glassy and hard carbons are clearly not amorphous. However the
electrical characteristics in some ways resemble those of amorphous
Ge and Si. Several authors [J2,3,9,13] have obtained measurements
which indicate that conductivity-temperature plots can be fitted
with the expression for variable range hopping (see Equation A-43
and discussion in Appendix D).

Data from a more recent study of Saxena and Bragg on the electrical conductivity [J9] and magnetoresistance [J10] of porous glassy
carbons are presented in Figures 36-38. The variation of the conductivity with temperature for two materials heat treated at 1400°C and

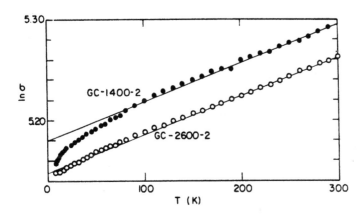

Figure 36. Plot of ln σ vs. T for 2 glassy carbons. (Figure from Saxena and Bragg [J9].)

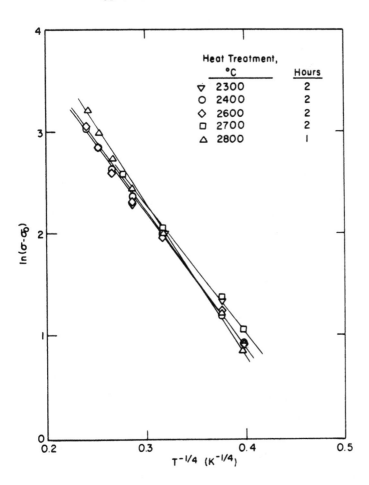

Figure 37. Plot of $\ln(\sigma - \sigma_0)$ vs. $T^{-1/4}$ for glassy carbon with high heat-treatment temperature. (Figure from [J9].)

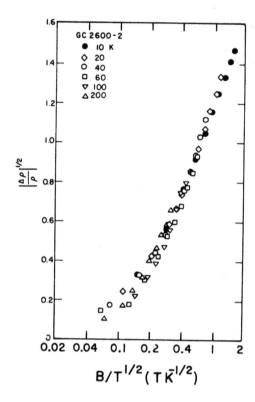

Figure 38. Magnetoresistance of a glassy carbon heat treated at 2600°C for 2 hr at various measurement temperatures. A universal curve is obtained when plotted against $B/T^{1/2}$. (Figure from [J10].)

2600°C are shown. Plots of $\ln \sigma$ vs. T were found to give straight lines over wide ranges of temperature. Samples heat treated below ~2000°C showed a sharp decrease below the linear curve at lower temperature.

Plots of $\ln(\sigma - \sigma_0)$ vs. $(1/T)^{1/4}$ are presented in Figure 37 for various values of HTT above 2000°C; $\ln\sigma_0$ is identified as the intercept of the straight line plot of Figure 36 and σ_0 is interpreted as the temperature-independent, boundary-scattering contribution to the conductivity. Further increase in σ was therefore ascribed to variable range hopping (see Appendix D) since $\ln (\sigma - \sigma_0)$ vs. $T^{-1/4}$ plots were straight lines. T_0 tends to higher values at lower values of HTT.

Conductivity of samples with HTT below ~2000°C was also found to follow the same behavior at higher temperature (the linear portion of the curve in Figure 36). The drop at lower temperature was ascribed to the *Kondo* effect, which is scattering from localized spins.

Saxena and Bragg also found a correlation for the negative magnetoresistance

$$\frac{\Delta\rho}{\rho_0} = f(\frac{B}{T^{1/2}})$$
(10)

This is illustrated in Figure 38 for HTT = 2600°C. Although a complete model could not be developed to explain these observations, the data were discussed in terms of a model in which localized spins formed a self-trap energy for hopping conduction. The negative magnetoresistance was then ascribed to the magnetic field-dependent scattering from these spins. As discussed in Appendix B.4, this is the model proposed by Toyozawa [A20]. Although his model may be applicable to certain carbons, it is not established that this is the mechanism responsible for negative magnetoresistance in glassy carbon. An alternative model has been proposed by Bright [ZD1] (see Appendix B.4).

IX. CARBON FIBERS

The feature that sets carbon-fibers apart from other carbons is the importance of the surface in the heat-treatment process. Typical carbon fibers have diameters ~5-10 μm, so that the surface can play a crucial role in determining free energies at different stages of carbonization and graphitization.

There has been tremendous scientific and technological interest in carbon fibers during the past few years because of the discovery of ways to prepare high strength materials. High strength fibers have been prepared from polyacrylonitrile (PAN) precursors in the U.K., from rayon-based precursors in the U.S., and were later prepared from pitch in Japan. The preparation procedure for PAN-based

fibers basically consists of preheating at ~220°C, followed by carbonization at ~1000°C, and finally a heat treatment between ~1500-
2500°C, sometimes under tension to increase alignment. Preparation
from other precursors involves similar carbonization and heat-
treatment processes.

Resulting fibers are sometimes divided into two classes, depending on whether heat treatment is above or below 2000°C. Type I are
heat treated above 2000°C and have high Young's modulus (~400 GPa or
~55 x 10^6 psi) and relatively low strength (~1.7 GPa or ~250,000 psi).
Type II are most commonly heat treated below 1500°C and have relatively lower modulus (~240 GPa or ~35 x 10^6 psi) and higher strength
(~2.8 GPa or ~400,000 psi). Commercial fibers are usually of Type II.

A review of the structure and physical properties of carbon
fibers has been given by Reynolds [K15] in Volume 11 of this series,
while Bacon [K2] has also contributed a chapter in Volume 9 on the
properties of fibers from rayon precursors.

The structure of carbon fibers has been the subject of intensive
research. Early work on high strength fibers established that carbon
planes were fairly well aligned along the axis of the fiber. Models
were developed in which the planes took the form of narrow ribbons.
However, more recent work has stressed the isometric nature of these
networks, as indicated in Figure 39. A longitudinal section taken
through the fiber shows the good alignment of the planes that undulate,
containing relatively straight segments with length L_a as high as
~100-200 Å (0.01-0.02 μm). However, a transverse section shows a
"thumbprint" arrangement of carbon networks.

Within a region of fairly straight carbon networks, the stacking
is regular for a characteristic distance L_c. Regions of this type
are known as fibrils, but it is again stressed that they are not
ribbonlike. Carbons networks extend over much greater distances
than L_a and L_c, with twists and bends. The growth of L_a and L_c with
HTT for PAN-based fibers is sketched in Figure 40, together with data
on the mean misorientation angle of carbon networks with respect to
the fiber axis, and the Young's modulus of elasticity (E).

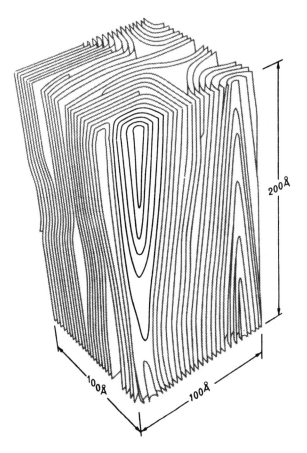

Figure 39. Model for the structure of a high strength carbon fiber.
(Figure from [K2].)

Graphite whiskers were earlier produced by Bacon [K1] who showed
that they possessed a rolled-up scroll form. These whiskers had
exceptionally high Young's modulus (~800 GPa ~100 x 10^6 psi) and
strength (~20 GPa ~3 x 10^6 psi). However, it has not been possible
to exploit their exceptional properties. The resistivity of such
whiskers as a function of temperature (Figure 2) is very similar to
that of a poor quality HOPG (Figure 13) or a pyrolytic carbon heat
treated to ~3000°C (Figure 22).

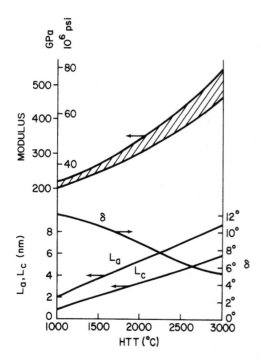

Figure 40. Sketch of the variation of turbostratic crystallite
dimensions L_a and L_c with HTT of the fiber. The mean crystallite
misorientation angle, δ, with the fiber axis and modulus of elasticity are included. The data are for PAN-based fibers but represent trends in fibers based on other precursors. (Figure based on
data from [K15].)

　　　　Curves of the variation of resistivity with temperature are
given in Figure 41. Throughout this section, resistivity will be
taken as measured along the axis of the fiber. The data in Figure
41 are for fibers prepared from mesophase pitch. Two different
structures are represented in Figure 41, *radial* and *random*. These
terms express the arrangement of carbon networks in a transverse
plane. Another possible arrangement is known as "*onion-skin*," in
which the networks are roughly segments of cylinders.

　　　　It is clear that fibers with random structure are less well
developed than those with radial. The increased stacking disorder in
the fibers with random structure prevents the growth of crystallites

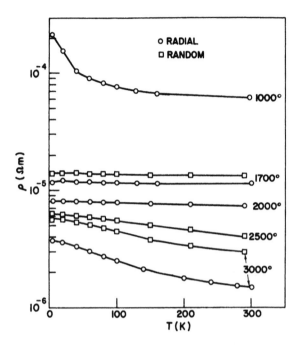

Figure 41. Resistivity as a function of temperature for fibers
prepared from mesophase pitch precursor, with both radial and
random structure. [Figure adapted from Bright and Singer [K5].)

in the a-direction. A further comparison can be made between these
data, and those for pyrocarbons in Figure 22. Fibers are structurally
less well developed than bulk material, as evidenced by resistivity
characteristics, and confirmed by x-ray and electron-microscope data.
This is related to the enhanced effects of the surfaces in fibers.

In comparing resistivity-temperature data in Figure 41 with
those in Figure 22 it may be concluded that defect scattering controls the mobility for all of the fibers shown throughout the temperature range up to ~300 K. The variation of resistivity with temperature is controlled by the density of carriers, with μ_c approximately constant for HTT $\gtrsim 2000°$C.

The temperature coefficient of the resistance also varies in a
smooth way with HTT. Usually the resistivity ratio $\rho_{77\ K}/\rho_{300\ K}$ is
used to express a measure of the temperature coefficient. Properties

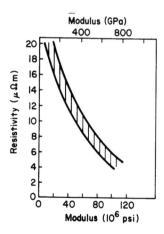

Figure 42. Sketch of the variation of fiber elastic modulus with resistivity. Data from fibers prepared from several precursors falls within the shaded area. (Figure adapted from [K8].)

of fibers with different precursors scale better against the electrical characteristics, such as $\rho_{300\ K}$ or $\rho_{77\ K}/\rho_{300\ K}$ than against HTT, and many correlations of properties are made against both of these properties.

For instance there is a strong correlation between the room temperature resistivity of carbon fibers and their modulus, which is in turn related to their strength. A summary graph [K8] of the resistivity vs. fiber modulus for fibers with several precursors is given in Figure 42. Figure 43 summarizes the correlation between the resistivity ratio $\rho_{77\ K}/\rho_{300\ K}$ and $\rho_{300\ K}$, and also with the x-ray crystallite size.

Until the highest heat-treatment temperatures are reached (~3000°C) there is little evidence for graphitization. The magnetoresistance at 1.4 T, plotted against resistivity ratio, and correlated with HTT is plotted in Figure 44. Bright [K4] has developed a semiquantitative model to explain the occurrence of negative magnetoresistance on the basis of the turbostratic structure. Positive magnetoresistance was related to the development of a graphitic structure. This topic is dealt with more fully in Appendix B.4.

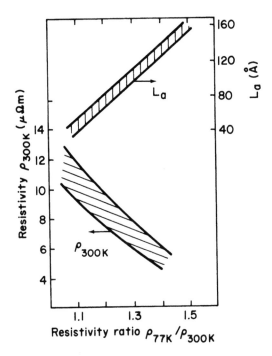

Figure 43. Sketch of the variation of the turbostratic crystallite dimension L_a and the room-temperature resistivity with resistivity ratio $\rho_{77\ K}/\rho_{300\ K}$. Cross-hatched areas represent the spread in values for fibers prepared from different precursors. (Figure adapted from [K17].)

Finally, a curve of the positive magnetoresistance vs. angle between the magnetic field and fiber axis is given in Figure 45 for B = 1.35 T [K17]. The curves compare two PAN-based fibers, one hot-stretched to increase crystallite ordering. Clearly the magneto-resistance anisotropy is lower for the stretched fiber. Robson et al. [K17] discussed the use of magnetoresistance anisotropy to evaluate the mean crystallite misorientation angle. Although de-tailed comparisons were not given with x-ray determination, they concluded the values based on magnetoresistance anisotropy were higher. However, it appears that an incorrect method of analysis was used, and the values appear reasonable (~15° for unstretched, ~6° for stretched fiber) if a correct analysis is used (see Appendix B).

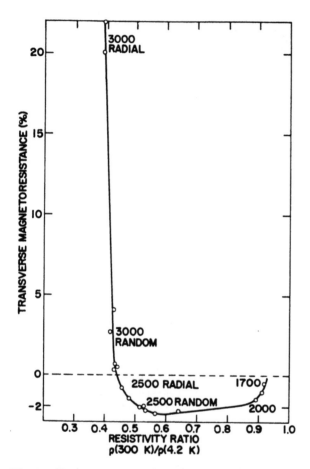

Figure 44. Variation of the magnetoresistance at 1.4T and 4.2 K.
Several fibers prepared from mesophase pitch precursor with both
radial and random structures are correlated against the resistivity
ratio $\rho_{300\ K}/\rho_{4.2\ K}$. (Figure from [K5].)

Recently there has been tremendous concern that air-borne carbon
fibers could have serious effects on the operation of electrical
machinery or microelectronic circuits. In the former case, electrical
power stations could conceivably be put out of operation by a saboteur
floating a quantity of fibers into the generator room. Also, the
electronic equipment in an aircraft could be compromised by a fire,
which could produce airborne fiber particles from the composite

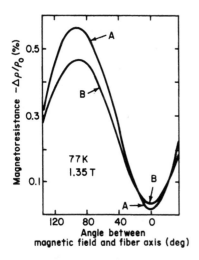

Figure 45. Variation of the magnetoresistance as a function of angle between B and the fiber axis at 1.35T, 7 K for PAN-based fibers. The curve A is for fibers hot-stretched by 22% for HTT of 2230°C. The curve B is for unstretched fiber with HTT of 2230°C. (Figure adapted from [K17].)

structure. Accordingly, attempts are being made to reduce the electrical conductivity of high strength fibers used in composites.

One approach is to introduce an impurity to modify the density of carriers. However, since commercial fibers have a turbostratic structure, simple calculations show that the conductivity will not be changed appreciably by changing ε_F near the band overlap region (see Appendix C). Based on analogy with boron-doped pyrolytic carbons, boron doping should *decrease* the resistivity as ε_F is depressed further (see Figure 50).

An interesting possibility would be to introduce a band gap into the electronic structure by doping with BN. A rough estimate for the band gap corresponding to a certain concentration of BN could be made by assuming a linear variation of band gap with composition between the zero gap of turbostratic (2D) graphite, and the 5 eV gap of pure BN. The technological difficulty associated with this approach is the introduction of nitrogen and boron into the lattice. It appears that nitrogen and boron are not stable in graphite at normal nitrogen and boron gas pressures.

While attempts are being made to reduce the conductivity of
fibers, other research is being conducted looking into the possibility
of increasing the conductivity, by intercalation, to a value close to
that of copper. The low-density, high-strength, high-conductivity,
intercalated fiber may have technological applications, and is considered further in Section XV-C.

X. CARBON FILMS

Highly disordered carbon films can be prepared in a number of different ways. Usually the film is vaporized onto a glass or oxide substrate by evaporation from a carbon source, which can be resistance
[L1] or electron-beam heated [L16,17]. Other techniques employed are
arc evaporation [L12] getter-sputtering [L10], or the glow discharge
of acetylene [L3].

Structural studies indicate that the films have short-range
order over distances ~10 Å, but no long-range order. Films with resistivities ~0.01-1 Ωm, have bond lengths characteristic of sp^2 bonding
but some authors argue that sp^3 bonds must also be present. Films
with a disordered diamondlike structure can also be prepared [L2,L7,
L8,L18]. The amount of impurity present depends on the preparation
conditions, particularly the ambient gas and its pressure, and the
rate of evaporation.

The room-temperature resistivities depend on the preparation
conditions (see Figure 1). Films prepared by several of the above
techniques have room temperature resistivity in the range ~0.01-1 Ωm,
for film thicknesses above 500 Å. Resistivities increase exponen-
tially with decreased thickness below ~500 Å, and the reason for this
is not completely understood. Attention will be focussed on films
with thicknesses greater than this limit, prepared by evaporation
techniques.

Conduction in these films is governed by a hopping process and
follows the variable-range Equation A-43 over a fairly wide range of
temperature. Figure 46 illustrates a plot of ln σ vs. $T^{-1/3}$ and

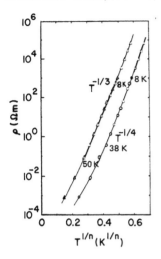

Figure 46. Plot of the ohmic (low current density) resistivity for carbon films as a function of $T^{-1/3}$ and $T^{-1/4}$. (Figure adapted from [L16].)

$T^{-1/4}$ for electron-beam evaporated carbon film in the temperature range 200-300 K. The data fit the $T^{-1/3}$ relationship better than the $T^{-1/4}$ one. According to Equation A-43 in Appendix D this indicates that the conduction is more nearly two-dimensional.

These data are taken in the low-current density regime, where Ohm's law is obeyed. At higher current densities nonohmic behavior is observed, with $j \propto E^{1/2}$ (Figure 47), where j is the current density (Am^{-2}) and E the electric field (Vm^{-1}). The data can be fitted to expressions of the Poole-Frenkel type:

$$j \propto \exp(e\beta E^{1/2}/kT) \tag{10}$$

where β is a constant, characteristic of the material. (A better fit was obtained by Morgan using a generalization of this equation.) In the high electric field region, higher current flows as a result of enhanced field emission--that is, a charge carrier in a potential well (see Figure A-17 in Appendix D) is thermally excited and then tunnels through a potential barrier whose height is reduced by the electric field.

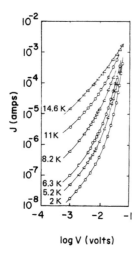

Figure 47. Plots of the current-voltage relationship for a film of
thickness 480 Å and an area 10^{-3} cm^2. (Figure adapted from [L16].)

        Carbon films have been prepared with a disordered diamondlike
form (see References, Section C). Resistivities can be very high
($\gtrsim 10^9$ Ωm), but depend sensitively on preparation technique. A variety
of measurement techniques, including optical, structural, resistance,
etc., indicate that tetrahedral (sp^3) bonding predominates in these
films and that the activation energy for conduction is ~1 eV. High
resistivities are also obtained by the glow discharge of acetylene
[L3] (~10^{10} Ωm) but in this case hydrogen impurity may be responsible
for the high activation energy for conduction (~1 eV).

XI. DOPED CARBONS AND GRAPHITES

The transport properties of graphite can be altered dramatically by
doping with foreign atoms. Boron is believed to act as a substitutional acceptor and silicon a neutral, substitutional impurity.
Donor impurities are usually in the form of metals that can form
lamellar compounds (see Section XV). After heating under vacuum
conditions, small concentrations remain at defect sites (residue

compounds). However, in high quality crystals, their concentration
can be reduced to the ppm level.

By doping with an acceptor or donor, the Fermi level can be
lowered or raised past the valence or conduction band levels, respectively. Since the density of electrons or holes in undoped graphite
is only ~$1/10^4$ atoms, only doping densities of ~100 ppm are required
to change graphite from a compensated (N = P) semimetal to a single-
carrier semimetal (the term *semimetal* is hardly appropriate here, but
a metal is usually associated with a Fermi level of ~0.2 eV at least).
During this doping process the magnitude of the Hall coefficient can
be changed as the value of (P - N) is altered by doping, and the
magnetoresistance is affected both by the change in the Fermi level
and the decrease of carrier mobility due to scattering by the ionized
dopants.

The effect of boron doping on the transport properties of NSCG
has been investigated by Soule [M14]. Boron can be diffused into
graphite by heat treatment above ~2500°C and, at least in part,
enters the graphite lattice substitutionally. It acts as an acceptor,
depressing the Fermi level. At about 50 ppm at OK, the Fermi level
is depressed below the conduction band edge so that the only carriers
are holes. At this doping level, however, both holes and electrons
are present at higher temperature (e.g., 300 K) due to thermal excitation of carriers into the conduction band.

The resistivity increases with boron doping since ionized boron
atoms act as scattering centers (Figure 48). This can be seen from
the reduction of mobility in Figure 48. Of greater interest is the
study Soule made of the variation of Hall coefficient with doping,
shown in Figure 49. Solid lines are theoretical curves for the change
in R_H at 0 and 298 K (see Appendix B). Note that the boron concentration at which R_H reaches a peak value shifts to a higher value at
298 K since this temperature is above the degeneracy temperature of
intrinsic graphite ($T_d = \varepsilon_f/k$). Also, at 298 K the peak value of R_H
is reduced because electrons are still excited when ε_f is in coinci-
dence with the band edge.

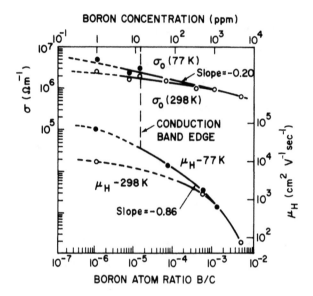

Figure 48. Curves of the variation of the conductivity and magneto-resistance (B = 0.394T) for boron-doped single crystals of graphite. (Figure adapted from [M14].)

In order to get good agreement between the calculated curve and experimental points, Soule had to assume 75% ionization efficiency for the boron atoms in the crystal. This could be interpreted as the percentage of B atoms in substitutional sites. In a review in Volume 7 of this series on doped carbons, Marchand [M9] gives a much lower value (5-20%) for polycrystalline carbons.

Klein also studied the electrical properties of boron-doped pyrolytic graphite and turbostratic graphite (Figure 50) [G15]. An increase of resistivity of pyrolytic graphite (3000°C) was found with boron doping similar to that for NCG while the resistivity decreased for as-deposited, boron-doped material.

The key to the different behavior of doped turbostratic material was given by the study of the variation of the Hall coefficient with temperature for different doping levels (Figure 51). Klein compared the experimental curves with those obtained for the simple two band model (Appendix B) with density of carriers obtained from the cylin-

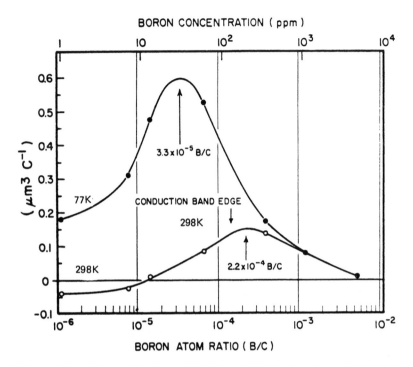

Figure 49. The variation of Hall coefficient extrapolated to B = 0
as a function of boron doping. Full lines are calculated values with
mobility ratio μ_n/μ_p = 1 and carrier density ratio N/P determined
from the band model, assuming 75% ionization efficiency. (Figure
adapted from [M14].)

drical band model (Appendix C). He showed that the as-deposited
(2100°C) turbostratic material was compatible with the model if the
Fermi level were depressed to -0.02 eV. In order to explain the
change from p- to n- type at ~500°C, the electron mobility was assumed
to be higher than the hole. Thus, as temperature increases and the
density of electrons increases, due to thermal excitation, then R_H
becomes progressively more negative. With boron doping, ε_f is progressively depressed, assuming a value of ~-1 eV at 1% boron (Figure
51).

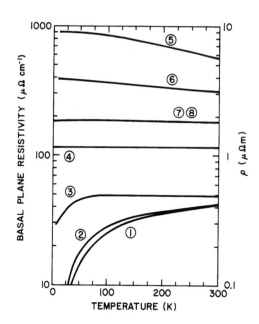

Figure 50.   Curves of the resistivity as a function of temperature
for boron-doped pyrolytic carbon and graphite: (1)-NSCG; (2) –heat-
treated PG (3600°C); (3) heat-treated PG (3000°C); (4) heat-treated
PG (3000°C) + 0.3 at. % boron; (5) as-deposited pyrolytic carbon
(2100°C); (6) as-deposited PC (2100°C) + 0.1 at. % boron; (7);
(8) PC + 0.3 and 0.6 at. % boron.   (Figure adapted from [G15].)

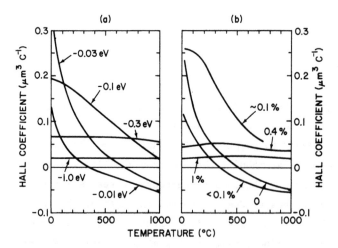

Figure 51.   (a) Curves of the calculated variation of the Hall
coefficient with temperature for turbostratic graphite using the
cylindrical band model.   Values of Fermi energy depression are
shown.   (b) Experimental data at .81T.   Boron doping levels are
shown.   (Figure adapted from [G15].)

Returning to the resistivity plots (Figure 50), the decrease of
resistivity upon boronation of turbostratic graphite can be directly
related to the increase in the total density of carriers with doping.
The mobility is expected to decrease, and Klein found that at ~0.3
at. % the resistivity decrease saturated. Presumably at a higher
doping level the increase in carrier density evidenced by Hall data
(Figure 51) is just offset by the decrease of mobility.

For boron-doped pyrolytic graphite, however, the carrier concentration is not expected to change appreciably until the Fermi level
is depressed to ~-0.1 eV (a further discussion will be given in
Appendix C). Accordingly, the predominant effect of boron doping for
this material is an increase of resistivity due to the decrease of
mobility, similar to that for NSCG (Figure 48).

A similar, anomalous variation of resistivity with boron doping
has been discussed for bonded polycrystalline graphite [M16]. Curves
showing the influence of boron doping on $\rho(T)$ are given in Figure 52.
It was remarked by these investigators that the resistivity was decreased by boron doping only at room temperature and below (cf.
Figure 50 for pyrocarbons), while at higher temperature the reverse
was true. These results can be rationalized when it is realized that
by 2000 K the scattering by ionized defects is small compared to
intrinsic (electron-phonon) scattering. Also, the thermal energy
kT ~ .17 eV, so that the density of carriers takes a value close to
the intrinsic value (at this temperature the fraction of ionized
donors, B^-, is probably small). The differences in resistivity above
2000°C only amount to ~5%. The lower resistivity of the sample without increased boron content in the temperature range between ~1000-
2000°C is probably attributable to the higher mobility. Again, it
is noted that differences are relatively small. At 1200°C, for example, resistivities of all samples lie within ~20% of each other
(Figure 52b).

A study has been made of the effect of silicon on the electrical
conductivity of pyrolytic deposits. Silicon was introduced by pyrolysis of propane gas mixed with $SiCl_4$. At high concentrations of sili-,

(a)

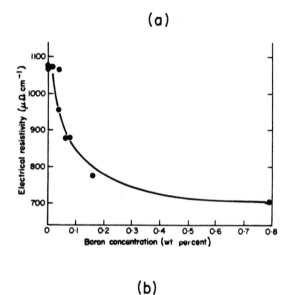

(b)

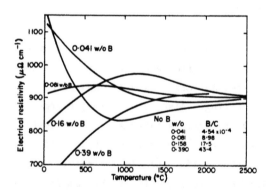

Figure 52. Effect of boron doping on the electrical resistivity of polycrystalline graphite. (a) Resistivity change with boron content at room temperature; (b) variation with temperature for samples with different boron levels. (Figures from [M16].)

con, precipitates of SiC were observed in the carbons. The results indicated that ρ_a was decreased, and ρ_c increased, with higher Si content. However, the main effect of the dopant was manifested

through induced structural modifications, particularly the higher
degree of orientation achieved for a given deposition temperature,
rather than a shift of Fermi energy.

Marchand has given a review of the properties of doped, heat-
treated, carbons in Volume 7 of the present series [M9]. Mrozowski
and co-workers [M2] have more recently performed some interesting
experiments on doped carbons in order to estimate the variation of
the band overlap with HTT. Using boron doping to introduce acceptors
and sodium to introduce donors, they swept the Fermi energy through
the band overlap region. From measurements of the variation of the
Hall coefficient they inferred the extent of the band overlap.

The relationship between band overlap and the variation of Hall
coefficient can be rationalized from Figure 53. For a given band
overlap the Hall coefficient will pass through a maximum and minimum
value as ε_F is changed. The amplitude of the changes in R_H (which
is denoted ΔR_H) depends on the band overlap ε_0. If the mobility
ratios remain fixed, the amplitude of ΔR_H quadruples for a decrease
in ε_0 by a factor of two. The curves in Figure 53 show the variation
of R_H at T = OK and 298 K. At higher temperature ΔR_H is reduced in
amplitude, and changes in R_H occur over a broader range of doping
levels because of the excitation of carriers with energy within the
range $\tilde{\varepsilon}_F \pm 2kT$.

Their experimental results showed that ΔR_H remained roughly
constant for HTT between 2400 and 3200°C. This is the range where
the degree of graphitization inferred from x-ray diffraction experiments changes markedly. Then for HTT between about 2000°C and 2400°C,
ΔR_H increased by about 40%, indicating a decrease of ε_0 of the order
of 20%. This range of HTT is where R_H goes through a maximum (Figure
29). For lower HTT's below the R_H maximum, ΔR_H decreased markedly.
This was interpreted as an increase in the band overlap. A further
result of this work was that, though electron carriers appear to be
more mobile for HTT above about 2200°C, hole carriers are more mobile
below.

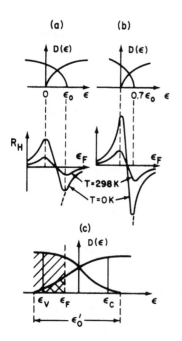

Figure 53. (a) Sketch of the variation of the Hall coefficient, R_H, with Fermi energy, ε_F; (b) same as (a), but band overlap is reduced to $0.7\varepsilon_0$, giving twice the amplitude of changes in R_H; (c) possible density-of-states curves for disordered carbon showing band overlap ε_0 due to disorder. (Figures (a) and (b) adapted from [M2].)

These results are surprising. In the HTT region above 2500°C, the band overlap is expected to decrease from the single-crystal value (~40 meV) to ~zero for turbostratic graphite.

One possible explanation for these results is that the amplitude of ΔR_H cannot be interpreted on the basis of a simple band model but, rather, on the basis of a partly disordered system. Accordingly, as the normal band overlap decreases, due to turbostratic stacking of planes, the bands are forming tails due to the increased disorder. This idea, sketched in Figure 53c, would qualitatively fit the data for ΔR_H. Also, holes would become more mobile than electrons at lower HTT, since the depression of the Fermi level would progressively excite the more mobile holes and fewer mobile electrons (see the curve

for the variation of carrier mobility in Figure A-16 of Appendix D).
However, the interpretation of these results must be regarded as pro-
visional, since (1) it is not known whether the amplitude of ΔR_H can
be directly related to the magnitude of the band overlap for a disordered system, and (2) the model for the disordered system discussed
in Appendix D has not been justified theoretically.

XII. NEUTRON-IRRADIATED MATERIALS

When graphites or carbons are irradiated with fast neutrons or electrons, atoms are knocked out of their lattice sites and Frenkel
defects are formed (vacancy and interstitial). The damage is accompanied by an increase in the basal resistivity and the c-axis spacing,
while the in-plane bond length decreases [N23]. Also, the c-axis
resistivity of oriented graphites decreases. The resistivity change
for several types of carbon and graphite is indicated in Figure 54.
A comparison of different measures of dose level is given in Ref.
ZN1.

   After irradiation the Hall coefficient becomes more positive and
the magnetoresistance is decreased (Figure 55). At low dose levels
the results for irradiated graphite are roughly compatible with a
model which assumes that the bands remain unchanged (rigid band model),
while the Fermi level is depressed by acceptors, creating a surplus of
holes. Electrons are probably localized at vacancies. The mobility
is decreased by the extra defect scattering, so that the magnetoresistance decreases. The curves of the change of Hall coefficient with
dose level are similar to those obtained as a function of doping level,
given in Figure 49. Note the similarity between the curves for graphitic materials, while the maximum in the curve for the pyrocarbon
occurs at lower dose level, but its peak value is lower. According
to the criterion illustrated in Figure 53 this would be characteristic
of a higher value of the band overlap, possibly caused by the "tail-
ing" of the conduction and valence bands.

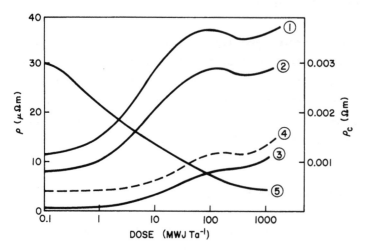

Figure 54. The increase of resistivity with neutron irradiation dose at 35°C for carbons and graphites: 1) polycrystalline graphite (high resistance direction); 2) polycrystalline graphite (low resistance direction); 3) pyrolytic graphite (PGCCL in Figure 22); 4) pyrocarbon; 5) same sample as (3) but c-axis resistivity (right-hand scale). (Figure adapted from [N22].) Abbreviations: 1 MWJ Ta^{-1} ≡ 1 megawatt day per tonne of adjacent material; 524 MWJ/Ta is roughly equivalent to 10^{26} neutrons of all energies per m^2. See [ZN1] for a discussion.

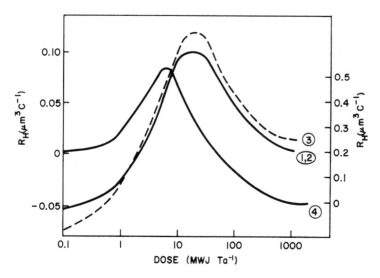

Figure 55. Variation of the Hall coefficient at 1.5T for the same samples as in Figure 54. The scale at the right of the figure is for pyrocarbon sample 4. (Figure from [N22].)

If the irradiation is carried out at low temperature and the
sample subsequently warmed, further changes occur in the conductivity
and galvanomagnetic properties, which are associated with the diffusion of defects into new positions and, eventually, their annihilation.
If the sample temperature is increased stepwise the changes occur over
a period of time after each temperature increment. In some instances
resistivity measurements can be useful in determining the activation
energy for the rate-limiting step in the process. After sufficient
time has elapsed the changes in electronic properties become very
slow, allowing a picture to be drawn of the change of the resistivity
with annealing temperature. (Remarks similar to those already made in
Section III concerning heat treatment of carbons need to be made about
the temperature-time dependence of annealing effects. Residence time
at the annealing temperature is an important variable.)

The most sensitive technique for observing changes in properties
with annealing would be to anneal at each temperature for a sufficient
length of time, then return the specimen to low temperature (e.g.
4.2 K) where defect scattering predominates. In this way the intrinsic scattering processes would be minimized and the effect of defects
maximized. The resulting curve of resistivity vs. annealing temperature is known as an isochronal annealing curve. This is the procedure
adopted by Iwata et al. [N11], whose results are illustrated in Figure
56a,b for annealing temperatures up to ~80 K, following irradiation
at 5 K. Surprisingly, annealing effects can be seen even below 10 K.
By differentiating the resistivity curve, annealing peaks can be
thrown into relief. One peak occurs at ~10 K, and another at ~75 K.

Both of the above peaks are associated with a decreased resistance. Higher annealing temperatures lead, however, to an increase
in resistance, first observed by Austerman and Hove [N1]. This anti-
annealing peak is maximum at ~95 K (Figure 57). A study of the electrical resistivity, magnetoresistance and Hall effect after irradiation at ~79 K showed that annealing in the temperature range 80-110 K
has the same effect on all three properties as an increase in irradiation dose (Figure 57). It was conjectured that the peak in the resis-

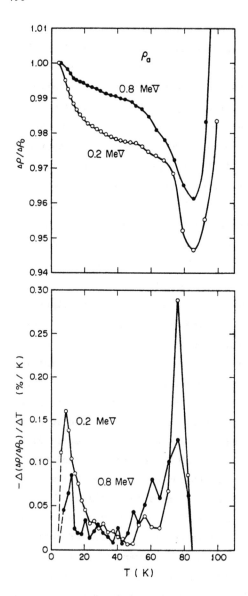

Figure 56. Annealing curves for radiation-damaged graphite:
(a) irradiated at 5 K, resistivity measured at 4.2 K [N11];
(b) irradiated at 5 K, data presented in differential form [N11].
(Figure from [N11].)

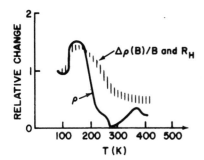

Figure 57. Illustration of the anti-annealing changes in resistivity, ρ, magnetoresistance, $\Delta\rho(B)/\rho_0$, and Hall coefficient R_H observed near ~120 K for polycrystalline graphite irradiated at 79 K. Relative changes are normalized to changes induced by the initial irradiation. At 270 K the resistance has fallen to its preirradiation value. (Figure adapted from [N24].)

tivity is associated with the motion of interstitials away from vacancies because of mutual repulsion. The separated interstitial with positive charge, and vacancy with negative charge, were then assumed to scatter electrons more effectively. However, the change in Hall coefficient was ascribed to the further trapping of charge by some defects, forming a doubly charged vacancy, for example.

Iwata et al. [N11] conjectured that the relatively small annealing peak at ~10 K was also due to motion of interstitials with activation energy ~0.02 eV. It is probable that the interaction of point defects with other defects, such as dislocations, needs to be considered before a complete model for the annealing processes can be established.

Other resistivity changes occur at the higher annealing temperatures associated with the motion of interstitials to form di-interstitials and eventually platelets of interstitials. At sufficiently high temperature the vacancies become mobile, and this probably gives rise to the annealing peak at ~1400°C for heavily damaged material in Figure 58.

If samples are irradiated at room temperature or above they can be subsequently cooled and rewarmed to room temperature without per-

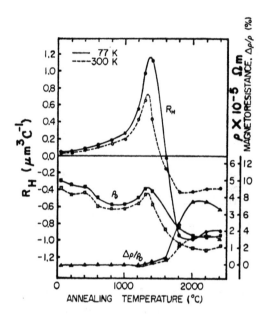

Figure 58. The resistivity measured at 77 K and 300 K for heavily
irradiated nuclear graphite annealed stepwise. (Figure from [N25].)

ceptible annealing effects in the electronic properties. Curves of
the change of resistivity with temperature for lightly irradiated
samples of highly oriented pyrolytic graphite are indicated in Figure
59. The progression of resistivity behavior with increase of dose is
very similar to that observed for pyrolytic carbons as the crystallite
size decreases (see Figure 22).

In order to assess changes in the carrier densities and mobilities
previous workers have used galvanomagnetic properties measured at relatively low magnetic field values (e.g. ~.5 T). However, it is well
established that galvanomagnetic data cannot be interpreted in this
field range to give reliable estimates for these parameters. Accordingly, a recent study was made of the galvanomagnetic properties of
HOPG irradiated to neutron doses in the region 0-3 x 10[17] neutrons
m[-2] in fields up to ~9 T. Galvanomagnetic properties were then
measured at low temperature where defect scattering predominates.

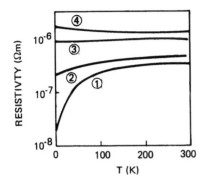

Figure 59. Resistivity as a function of temperature for samples of HOPG irradiated to the following fast neutron fluxes (Energy > 1 Mev): 1) unirradiated; 2) 0.18 x 10^{17} nvt; 3) 0.72 x 10^{17} nvt; 4) 3 x 10^{17} nvt. (Data from [N7].)

This work conclusively established that the rigid-band model was not compatible with the experimental results. It is not known whether the problem lies with the present state of knowledge of the electronic energy bands or whether other factors such as electron-hole coupling are being neglected.

XIII. C-AXIS CONDUCTIVITY

Typical ρ(T) curves for c-axis conductivity are given in Figure 60 and anisotropies are compared in Table 3. A major difference is to be noted between the anistropy ratios for HOPG and similar materials (~5000 at 300 K, rising to ~10^5 at ~4 K for some specimens) and NSCG or KG (~100 at all temperatures). It is to be noted that high anistropy values have been reported for several specimens of purified natural graphite [E8] but the specimens have not been single crystals, or allowed measurements to be made over individual crystallites of twinned crystals.

Neutron and electron irradiation lowers the c-axis resistivity of HOPG [F8] and pyrolytic carbons [G3] but increases that of NSCG [O2]. Boron doping decreases the c-axis resistivity of pyrolytic carbons (Figure 61) [G15].

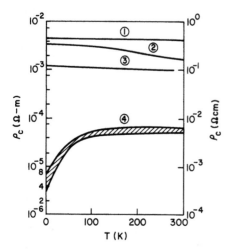

Figure 60. C-axis resistivity as a function of temperature for
several types of graphite. (1) as-deposited pyrolytic carbon (2100°C)
[G13]; (2) heat-treated pyrolytic graphite (3000°C) [G13] and HOPG
[F8]; (3) neutron-irradiated HOPG (~10^{18} nvt. E > 1 Mev) [F8];
(4) NSCG and KG [E12] and [012].

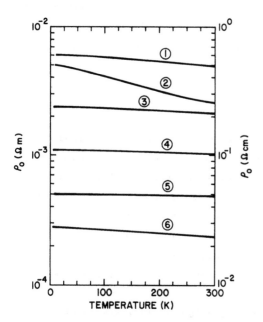

Figure 61. Variation of c-axis resistivity for boron-doped pyrolytic
carbons: 1) as-deposited pyrolytic carbon (2100°C); 2) heat-treated
pyrolytic graphite (3000°C); 3) as-deposited + 0.1 at. % B; 4) as-
deposited + 0.3 at. %; 5) as-deposited + 0.6 at. % B; 6) 0.3% B heat-
treated to 3000°C. (Figure adapted from [G15].)

Table 3. Summary of Experimental Data for C Axis Resistivities of Graphite

Material	Authors	T(K)	$\dfrac{\sigma_c}{\Omega m^{-1}}$	$\dfrac{\sigma_a}{\sigma_c}$	$\dfrac{\sigma_c \, 4.2\ K}{\sigma_c \, 300\ K}$
Natural graphite (Ceylon, Mexico)	Washburn [E18]	300	8.3×10^4	100	--
Natural graphite (Ceylon)	Dutta [E8]	300	10^4	104	--
Natural graphite (Ticonderoga)	Primak, Fuchs [E12]	300 / 77 / 4.2	$1.5\text{-}2.3 \times 10^4$ / 3×10^4 / 4×10^6	110-170 / 150 / 120	19
Natural graphite (Ticonderoga)	Primak [O7]	300	2×10^4	130	--
Natural graphite (Ceylon)	Bhattacharya [E7]	300	$0.7\text{-}2.5$	$10^4\text{-}10^5$	--
Natural graphite (Ticonderoga)	Wagoner [O14]	300	3.3×10^4	80	--
Kish graphite	Tsang, Dresselhaus [O12]	300 / 4.2	$1.3\text{-}1.5 \times 10^4$ / $11.5\text{-}32.7 \times 10^6$	--	8.8-22
Pyrolytic carbon ($T_d = 2200°C$)	Blackman, Saunders, Ubbelohde [G3]	300	125	5500	--
Pyrolytic carbon ($T_d = 2500°C$)	Klein [G13]	300 / 4.2	83 / 55	5000 / 3500	0.66
Pyrolytic graphite (HTT = 3000°C)	Klein [G13]	300 / 4.2	385 / 200	5200 / 1.6×10^4	0.52
HOPG (annealed 3500°C)	Spain, Ubbelohde, Young [F8]	300 / 4.2	590 / 380	3800 / 8.8×10^4	0.65

It is to be noted that the four-point technique for measuring
resistivity (see Appendix A-4) will tend to underestimate the resistivity ratio. Accordingly the higher values reported for HOPG using
this technique cannot be explained as a measurement error. Since
c-axis samples are generally small, 2-point techniques have sometimes
been used, giving enhanced values of ρ_c due to contact resistance.

When current flows parallel to the c-axis, the magnetoresistance
of HOPG is anomalous [F8] (no measurements on single crystals have been
reported, only measurements on single crystal aggregates [E6]). The lon-
gitudinal magnetoresistance $(\overline{J11}\overline{C11}\overline{B})$ is very much larger than the transverse $(\overline{J11}\overline{C1}\overline{B})$. A typical rotation diagram is illustrated in Figure 62.
Apparently, the small transverse magnetoresistance can be explained as
a spurious effect from the misorientation of the crystallites:

$$\left(\frac{\Delta\rho}{\rho_0}\right)_T (B_T) \sim \left(\frac{\Delta\rho}{\rho_0}\right)_L (B_L < \cos\delta >) \tag{12}$$

It is interesting to note that in very high magnetic fields
($\gtrsim 10$ T) the longitudinal magnetoresistance may saturate, or even
decrease [T3]. Accordingly the anisotropy ratio may decrease to
as low as ~50, from initial values at B = 0, T = 4.2 K approaching
~10^5 [T3].

The pressure coefficient of c-axis resistance is approximately
the same for HOPG [S13] pyrolytic graphite [O5] and single-crystal
aggregates [O5] ($d\ln\sigma_c/dp = -28 \times 10^{-6}$ bar^{-1} at room temperature)
and is roughly independent of pressure. Small ($\lesssim 0.1\%$ of R_0) irrepro-
ducibilities can be noticed at pressures less than ~100 bars for HOPG
similar to those observed in the elastic behavior [F3]. The above
value of $d\ln\sigma/dp$ is very close to the experimental value of the
pressure dependence of the density of carriers $d\ln (N + P)/dp$ [A18].

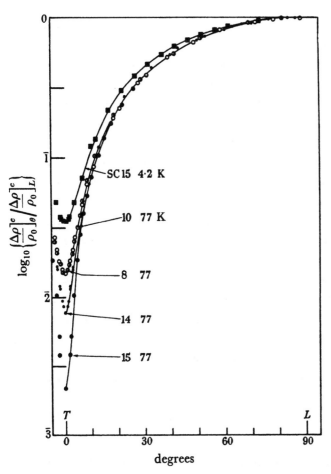

Figure 62. Magnetoresistance rotation diagrams for several specimens
of HOPG at 77 and 4.2 K. (Figure from [F8].)

A. Intrinsic Resistivity: The Case for NSCG and KG

At the present time there is no satisfactory explanation for the
c-axis conductivity in these materials. There are two basic possibilities, the first would be that the resistivity observed in NSCG
and KG is intrinsic; and therefore the higher anisotropy ratios
observed in HOPG would be due to defects in these materials. Two
different explanations for this behavior (lower resistivity in NSCG
and KG, and higher anisotrophy ratios in HOPG) are possible:

   1. The first begins with the observation that high resistivity
arises from thin sheets of disordered material separating crystalline
material. These disordered sheets effectively control the resistance
to current flow in the c direction. Recent scanning electron microscope studies of carefully prepared basal surfaces of HOPG samples
have revealed the presence of thin sheets with contrasts differing
from those of the parent material [Q10]. These sheets were interpreted as microcracks of thickness ~0.02 μm, separated by ~2 μm. In
order to explain the observed resistivity of HOPG a simple calculation shows that the resistivity of these regions (if assumed composed
of highly resistive material, such as amorphous carbon) must be as
high as ~1 Ωm. In this case the relative resistance to current flow
offered by the crystalline regions is clearly very much smaller than
that of the high resistivity regions or cracks.

   2. Alternatively, the high resistance must arise from a highly
tortuous current path along basal directions, around cracks.
Neither of these possibilities is consistent with data for HOPG for
three reasons:

   1. The highly reproducible pressure coefficient of resistance
      for different samples, or upon pressure cycling.

   2. The longitudinal magnetoresistance which is large, but not
      proportional to the basal magnetoresistance.

   3. The temperature dependence of the resistivity.

Yet a third possibility is that high c-axis resistivity in HOPG
arises from the localization of current carriers by defects, such as

arrays of basal dislocations [06]. It is well known that the predominant basal dislocations in graphite split into partial dislocations. The region between the partial dislocations is a stacking
fault with the ...ABC... stacking sequence of rhombohedral graphite.
This stacking fault can act as a reflecting barrier for electrons.
At sufficiently high dislocation density the carriers could become
localized by these stacking faults, by a similar mechanism to that
in isotropic, disordered materials (see Appendix D).

A calculation [06] which treated the stacking faults as reflecting barriers for the electrons showed that a resistivity ratio of 10^4
could be obtained at room temperature if the distance between stacking
faults was 200 Å. This corresponds to a dislocation density of
2.5×10^{15} m^{-2}. Crystals of HOPG typically have densities ~3.10^{13} m^{-2}
NSCG ~2.10^{10} m^{-2}. As can be seen from Figure 63, the model predicts
the wrong temperature dependence for the resistivity.

The influence of stacking disorder on the conductivity of GaSe
and similar layer materials has recently been considered [A11]. It
was concluded that Anderson-Mott type localization (see Appendix D)
would occur for

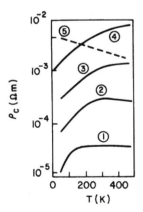

Figure 63. The calculated c-axis resistivity for different values of
the effective distance between stacking faults. 1) ideal crystal
($\ell = \infty$); 2) $\ell = 1$ μm (10,000 Å); 3) $\ell = 0.1\mu$ m (1000 Å); 4) $\ell = 0.01$ μm
(100 Å); 5) experimental data for HOPG or pyrolytic graphite ($\geq 3000°C$).
(Figure adapted from [06].)

high concentrations of stacking disorder. However the model was not
able to make a clear prediction of the minimum concentration of
defects necessary to create localization.

B. Intrinsic Resistivity: The Case for HOPG

An alternative approach in seeking an explanation for the c-axis conductivity is to assume that the resistivity values observed in HOPG
are more nearly representative of intrinsic behavior, and that lower
values of NSCG and KG are due to defects which "short" the intrinsic
resistance. Some credence for this can be derived from the observation that the resistivity ratio for NSCG is independent of temperature,
within experimental uncertainty.

It is possible that the lower resistivity observed in single
crystal graphite could result from the presence of screw dislocations
with Burgers' vector parallel to $\bar{c}$. Thus the lower resistivity would
be produced by the passage of current along basal paths in a spiraling
motion about the dislocation core. If there are N-dislocations per
unit area with Burger's vector $c/2$, and interplanar conduction is
neglected apart from this defect contribution, then the conductivity
anisotropy could be written approximately as:

$$\frac{\rho_c}{\rho_a} = \frac{8\pi}{N' c^2 \ln(1/2\ a\sqrt{N}\ )} \tag{13}$$

If reasonable values are used for the dislocation core radius, a,
(of the order of 0.1-10 Å) then a very high dislocation density N'
$\sim 10^{19}\ m^{-2}$ is required to yield the observed ratio of ~ 100 for NSCG
or KG. Experimentally the dislocation density is of the order of
$10^{10}\ m^{-2}$ [B1] for these materials.

If a reduction in c-axis resistivity results from basal shorting,
then paths on a macroscopic scale are required, rather than on a
microscopic. A simple example is afforded by the case of a sample

consisting of two crystals with c-axes inclined at an angle to each
other [E12]. Quite clearly the measurement of c-axis resistivity in
a small crystal of SCG must be accompanied by a detailed structural
investigation, both before and after the measurement. This was essentially the procedure adopted by Primak [07] and Primak and Fuchs [E12],
who used techniques to measure ρ_c for single-crystal regions of twinned
crystals.

Klein [G15] developed a simple model for the change in c-axis
resistivity with degree of preferred orientation. If the normalized
density of crystallites with angle δ to the principal, or average,
c-axis is $n(\delta)$, then, for cylindrical symmetry, he assumed

$$n(\delta) = \cos^m \delta \qquad (14)$$

If the conductivities within each crystallite were σ_a and σ_c, then
for a sample without pores the bulk conductivities would be; with
reference to Figure 64

$$\sigma_{||} = \sigma_a \left[1 - \left(\frac{1 - \sigma_c}{\sigma_a} \right) < \sin^2 \delta \cos^2 \phi > \right]$$

$$\sigma_{\perp} = \sigma_c \left[1 - \left(\frac{1 - \sigma_c}{\sigma_a} \right) < \cos^2 \delta > \right] \qquad (15)$$

Then, suitably averaging, using Equation 14 gives

$$\frac{\sigma_{11}}{\sigma_c} = \frac{(m + 2)\sigma_a/\sigma_c + 1}{(m + 1) + 2\sigma_a/\sigma_c} \qquad (16)$$

Graphs of $\sigma_{11}/\sigma_{\perp}$ as a function of σ_a/σ_c are given in Figure 64.
The value m = 8000 roughly corresponds to a highly graphitized pyro-
lytic sample with $< \delta > \sim 0.5°$, while m = 80 is characteristic of a
heat-treated ($\sim 2800°C$) pyrolytic carbon with $< \delta > \sim 10°$, and m = 8
an as-deposited ($T_d \sim 2100°C$) sample (see Figure 20). Note that if

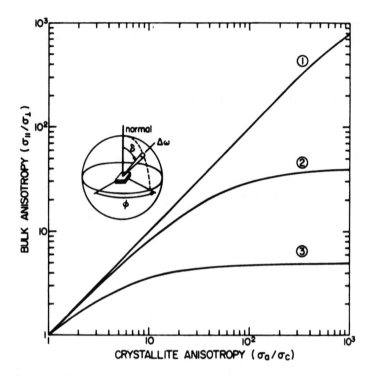

Figure 64. Graph of the bulk anisotropy ratio as a function of crystallite anisotropy ratio for three values of the misorientation index, m: 1) m = 8000; 2) m = 80; 3) m = 8. (Figure adapted from [G15].)

m = 0, the material is isotropic in its bulk resistivity. The effect of crystallite misorientation on the bulk anisotropy ratio is only slight for $\sigma_a/\sigma_c \lesssim 10^3$, if m $\gtrsim$ 5000, but crystallite misorientation should reduce the bulk anisotropy severely for m = 8. As seen from Figure 61, the bulk resistivity of an as-deposited pyrocarbon (T_d = 2100°C, m ~ 8) is high (~5 x 10^{-3} Ωm) so that bulk anisotropy is high (>10^3). This was interpreted by Klein [G15] to signify that microcracks were responsible for the high c-axis resistivity in pyrocarbons. However, the same arguments cannot be used for HOPG, since $\delta \lesssim 0.5°$ (m > 8000).

XIV. DIAMOND

Pure diamond would be an insulator in which electrical conduction
occurs by the excitation of electrons across the band gap, leaving
hole states in the valence bands (see Figure 4). If the band gap
energy is ε_g, and the effective masses of holes and electrons are
m_p^* and m_n^* respectively, then the densities of holes P and electrons
N is given by an expression of the form

$$NP = 4 \left(\frac{2\pi kT}{h^2}\right)^3 (m_n^* \, m_p^*)^{3/2} \exp^{-\varepsilon_g/kT} \tag{17}$$

$$N = P$$

If $m_{n,p}^* \sim 0.25 \, m_0$, where m_0 is the free electron mass (9.11×10^{-31} kg),
and $\varepsilon_g = 5.3$ eV, then the carrier densities for pure diamond would be
extremely small (N = P = 6×10^{-22} m^3 at 300 K) and resistivity too
high to measure. Nonequilibrium populations of electrons and holes
could be created by illumination with ultraviolet radiation (photon
energy $>\varepsilon_g$).

However, impurities control the density of carriers by providing
donor or acceptor levels within the band gap. Most naturally occurring diamonds still have extremely high resistivity values (e.g.
$\sim 10^{12}$ Ωm). However, a few natural diamonds, which are usually light
blue in color, have resistivity $\sim 10^2$ Ωm, characteristic of semiconductors. They are classified as type IIb diamonds (see Refs. P8 and
P12), and often referred to as semiconducting diamonds.

The conductivity of semiconducting diamonds is usually controlled
by Al acceptors lying ~ 0.38 eV above the valence band. A typical
resistivity plot is given in Figure 65. In the region below ~ 300 K
($1/T \gtrsim 0.003$ K^{-1}) the slope of the log ρ vs. $1/T$ plot is characteristic of excitation of electrons from the valence band into the
acceptor level (~ 0.38 eV). The minimum in the curve at T $\sim 600-700$ K
($1/T \sim 0.0015$) occurs because the acceptors states are saturated.
From the Hall curve in Figure 65 the acceptor density is $N_A \sim 2 \times 10^{22}$
m^{-3}, and the mobility at 300 K is ~ 0.1 m^2 V^{-1} sec^{-1}. In this

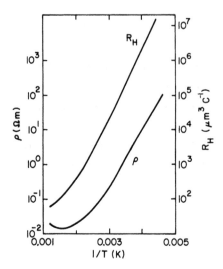

Figure 65. Typical variation of the Hall coefficient and resistivity
for semiconducting diamond. (Figure adapted from data in [P12].)

temperature region, $\mu_p \alpha T^{-3/2}$ characteristic of intrinsic scattering
by phonons. At lower temperature, scattering by ionized and neutral
impurities predominates.

Synthetic diamonds have been grown with several impurities such
as Be, B, Al, Ti (for a review of synthetic diamond growth, see Ref.
P5). Also, impurities can be diffused into natural or synthetic
crystals. In both these cases only p-type conductivity has been produced. However, donor impurities such as Li,P can produce n-type
conductivity when these elements are introduced by ion-implantation
techniques [P25,P30]. Point defects introduced by electron or neutron
irradiation can also produce semiconducting behavior [P10,P15,P16,P23].
With heavy irradiation the density can decrease by up to ~45%, and the
resistivity decrease to ~3 x 10^{-3} Ωm [P23].

XV.   INTERCALATION COMPOUNDS OF GRAPHITE

Graphite shares the ability to form intercalation compounds with a
number of other layer materials, such as MoS_2, $TaSe_2$, etc. However,

there are at least two features of these compounds that are unique to the graphite parent. The first is that graphite can form compounds with both acceptors and donors. The second is that several distinct compounds can be formed with the same intercalating species. Usually the most highly intercalated compound has a formula such as C_8X or C_6X. The distinct compounds formed with lower concentration of intercalant then have formulae such as C_8X, $C_{16}X$, $C_{24}X$ ... or C_6X, $C_{12}X$, $C_{18}X$ .... In such cases, at least for some compounds, the stoichiometry of the terminal compound is retained in fully intercalated layers, which are then separated by unintercalated graphite planes (Figure 66).

This property of intercalated compounds is sometimes referred to as *staging*. The compound with composition $C_{24}K$ is a *third stage* compound, since three graphite layers exist between each layer of intercalant. The chemical composition could be written $C_{8R}K$, where R is the stage. Although compounds undoubtedly exist where the insertion of layers is at least partly random, the concept of stage is useful, even for dilute compounds.

Stage 1 KC_8	Stage 2 KC_{24}	Stage 3 KC_{36}	Stage 4 KC_{48}	Stage 5 KC_{60}
oo oo oo A	oo oo oo A	oo oo oo A	oo oo oo A	oo oo oo A
--------	--------	--------	--------	--------
oo oo oo A	oo oo oo A	oo oo oo A	oo oo oo A	oo oo oo A
--------	oo oo o B	oo oo o B	oo oo o B	oo oo o B
oo oo oo A	--------	oo oo oo A	oo oo oo A	oo oo oo A
--------	oo oo o B	--------	oo oo o B	oo oo o B
oo oo oo A	oo oo oo A	oo oo oo A	--------	oo oo oo A
--------	--------	oo oo o B	oo oo o B	--------
oo oo oo A	oo oo oo A	oo oo oo A	oo oo oo A	oo oo oo A

oo oo oo	CARBON LAYER
--------	POTASSIUM LAYER

Figure 66. Illustration of the concept of *stage* for graphite intercalation compounds. The compound sequence, $C_{8R}K$, is used and the stacking sequence of planes indicated. (Figure from L. Vogel.)

Graphite intercalation compounds have been studied for a long period of time, but Ubbelohde[*] was the first person to recognize their potential as synthetic metals [Q67]. However, the recent discovery of room temperature conductivity as high as copper[†] in certain strong acid compounds, such as SbF_5, AsF_5 [Q81], has spurred a vigorous research effort into the properties of these materials. Several reviews have appeared recently [A3,A6,Q25,Q27]. A comparison of principal conductivites for several compounds is given in Table 4.

The change of conductivity with intercalation is dramatic. Initially, as the sample takes up the first quantities of intercalant, the conductivity increases quite steeply with weight uptake. The

---

[*]A. R. Ubbelohde coined the phrase "synthetic metal" in 1959.
[†]Recent data presented at the Second International Conference on Intercalation Compounds of Graphite, to be published in *Synthetic Metal* (F. L. Vogel, ed.), suggest that the conductivity may be only half that of copper.

Table 4.   Comparison of Principal Conductivities
           for Graphite and Selected Compounds

	σ_a $(10^6 \ \Omega^{-1} \ m^{-1})$		σ_a/σ_c	$\dfrac{\alpha_a \ 4.2 \ K}{\alpha_a \ 300 \ K}$
	4-point	Contactless technique		
HOPG	2.5	2.6	2.3×10^3	4-18
RbC_8	10	9.1	120	--
KC_8	10	12	30	220
$C_{16}Br_2$	22	20	7×10^4	--
$C_{16}ICl$	35	37	1.5×10^4	--
C_6HNO_3	25	13	2.0×10^5	--
$C_{12}HNO_3$	25	40	--	--
C_8AsF_5	25	47	$>10^6$	5.5-12.7
$C_{16}AsF_5$	22	58	$>10^6$	4.7
$C_{12}SbF_5$	21	50		

Source: Data selected from Fischer [Q25].

weight uptake is accompanied by a swelling associated with the insertion of the intercalant between the layers, pushing planes apart. At
high-weight uptakes the increase in conductivity tends to level off,
and in some cases maximum conductivity is found for second or third
stage compounds (Figure 67).

When intercalation takes place in a perfect crystal of graphite,
all atoms are located between layer planes. A lamellar compound is
formed. When the graphite is imperfect, some intercalant is located
in sites associated with defects, and is much more strongly bonded.
Thus, after a lamellar compound has been heated to remove intercalant,
some remains in the material. If no lamellar compound phase remains,
the remaining material is said to be a *residue compound*. A compound
formed from a carbon will generally contain both lamellar and residue
compound. The term residue compound is sometimes, erroneously, associated with the partly lamellar, partly residue compound formed by
treating a compound to a reduced partial pressure condition at room
temperature which fails to remove all the lamellar phase. This misuse
of the term is to be guarded against when reading the literature.

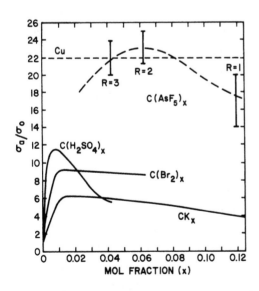

Figure 67. Increase in the basal conductivity of intercalated compounds of graphite plotted as a function of mole fraction of intercalant. (Data from [Q25] and [Q84].)

The arrangement of atoms within a layer may be ordered or disordered. A transition from an ordered to a disordered arrangement
can be inferred from structural studies, and has an important effect
on the conductivity. The disorder in many ways is analogous to a two-
dimensional melting of the intercalant within a layer, and is associated with anomalies in thermodynamic parameters such as the expansion coefficient (particularly α_c) and specific heat [A22].

The increase in conductivity of intercalated compounds over the
parent graphite undoubtedly stems from the increase of carrier density.
However, the high mobility of the carriers appears to be of major importance. Some representative data comparing average carrier densities and mobilities for a donor compound, $C_{36}Cs$; an acceptor compound,
$C_{16}(AsF_5)$; the parent graphite; and the metal copper are given in
Table 5. Compared to copper the carrier densities of the intercalated
compounds are low, but mobilities are high. The mobilities are relatively low, however, compared to the parent graphite (HOPG). The high
mobilities in intercalation compounds lead to certain practical con-
sequencies, which are discussed in Section C.

An enormous number of elements and compounds can be inserted
between the layer planes. Donor compounds, in which the Fermi energy
is raised, appear only to be formed with metal intercalants. Acceptor
compounds can be formed with elemental (e.g. Br_2) and compound intercalants (e.g. $FeCl_3$, HNO_3). Highest conductivities are found for
acceptor compounds. To date the second-stage acceptor compound
$C_{16}AsF_5$ has the highest measured room temperature conductivity
(Table 5). The reason for higher conductivity in the acceptor compounds has not been established.

In the following pages the conductivity of donor and acceptor
compounds will be discussed separately, and the effects of defects
in the parent carbon on conductivity are considered in Section XV.C.
Accurate data on transport properties of intercalation compounds are
subject to large uncertainties because of the high electrical anisotropy and the difficulty of maintaining good electrical contact while
the compound is swelling. Perhaps the only reliable measurements,

Table 5. Comparison of Carrier Densities and Mobilities
          (All Data at 300 K)

Substance	Cu	Graphite	$C_{36}Cs$ [a]	$C_{16}AsF_5$ [b]
$\sigma \ \Omega m^{-1}$	58×10^6	2.5×10^6	12×10^6	58×10^6
$N, P \ (m^{-3})$	$N = 8.5 \times 10^{28}$	$N+P = 2.2 \times 10^{25}$	$N \sim 1-2.10^{27}$	$P \sim 3 \times 10^{26}$
$\mu_c \ m^2 V^{-1} sec^{-1}$	4.3×10^{-3}	0.71	$\sim 4-8 \times 10^{-2}$	~ 0.2

[a]Values for $C_{36}Cs$ assume ~1 electron/Cs atom.
[b]Data for $C_{16}AsF_5$ based on magnetoresistance data of Ref. Q85.

particularly on acceptor compounds, are those obtained by a contactless eddy-current technique (see Ref. Q84 and Appendix A.4).

A. Donor Compounds

The increase in the conductivity of donor compounds can be understood
by considering the increase in the density of carriers. For weak compounds each intercalated metal atom may be considered to donate one
electron to the conduction band of graphite. This raises the Fermi
energy, reducing the density of holes, until only electrons are
present. In the density-of-states curve sketched in Figure 68, it
has been assumed that the density of states of graphite is unchanged
(rigid-band model) except for the addition of donor states associated
with metal atoms. If there are N intercalant atoms in the crystal
there will be 2N levels in this subsidiary band, and that N electrons
are donated from it to the graphite levels.

     For stronger compounds (e.g., first- and second-stage compounds)
the Fermi level is believed to be raised above the bottom of the
metallic density-of-states curve (see Figure 68). In this case less
than one electron per metal atom is transferred into the graphite
conduction band (the fractional ionization is less than unity).

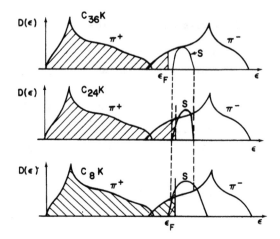

Figure 68. Hypothetical sketch of the density of electronic states
for $C_{36}K$, $C_{24}K$, C_8K.

In reality it is expected that the presence of the metal ions
between the graphite layers will modify the energy bands of the
graphite π-electrons. However, energy band calculations carried out
for the first stage potassium graphite compound (C_8K) [Q42] indicate
that the above model is at least reasonable from a qualitative point
of view. A sketch of the calculated Fermi surface is given in Figure
69. The nearly spherical piece of the Fermi surface corresponds to
the metallic electrons, while the nearly cylindrical pieces at the
corners of the Brillouin zone correspond to the graphite π-electron
states.

A comparison of d.c. conductivity and optical reflectance data
indicates that the metallic electrons have a much shorter relaxation
time than the π-electrons for basal conduction, so that π-electrons
dominate in basal conductivity. However, metallic electrons dominate
the c direction conductivity since their effective mass component is
much smaller in the c direction than for π-electrons.

The conductivity is expected to increase as the concentration
of metal atoms increases, mainly due to the corresponding increase in
the density of carriers. The mobility is expected to decrease, but

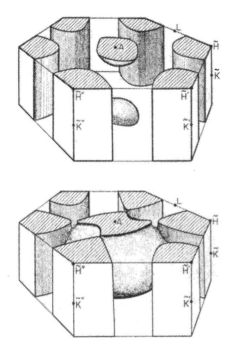

Figure 69.   Fermi surfaces for C_8K.   (Figure from [Q42].)

at a slower rate. If electron-phonon scattering predominates, simple
theory predicts that the relaxation time depends on the inverse of
the density of states at the Fermi energy, $D(\varepsilon_f)$. Also, the effective
mass is expected to increase as the Fermi level rises. Both of these
effects will contribute to the decrease in mobility (see Equation 4).

Once the Fermi level is raised sufficiently, so that only electrons are present (this occurs for a very weak compound), a single-
band model is expected to give a good representation of the galvanomagnetic behavior. If the fractional ionization is unity, then the
Hall coefficient should be $\sim R_H = -1/NE$, where N is the number of
metal atoms per unit volume. A sketch showing the variation of Hall
coefficient for several donor compounds of graphite at 4.2, 77, and
300 K is given in Figure 70. Markers are included to indicate the

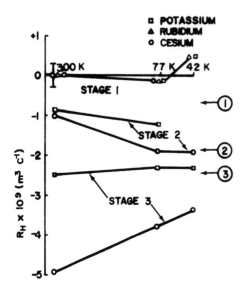

Figure 70. Variation of the Hall coefficient at ~300 K, 77 K and 4.2 K for alkali metal compounds of graphite. (1), (2) and (3) refer to values expected for stage 1, 2, 3 compounds, respectively, on the basis of a single-band model with one donated electron per metal atom. (Figure from [Q25].)

expected value for the Hall coefficient for first-, second-, and third-stage compounds based on the assumption of unit fractional ionization. The data do not fit this simple model except for the third-stage compound of potassium graphite ($C_{24}K$) and second-stage cesium graphite at lower temperature. The Hall coefficient even changes sign for the first-stage compounds between 77 and 4.2 K.

If the data for the first-stage compounds were analyzed on the basis of a single-band model ($R_H = -1/Ne$) the charge transfer would fluctuate between several electrons *donated* per atom at 300 and 77 K, to several electrons *accepted* at 4.2 K! Quite clearly the single-band model is not applicable. Also, a model with two electron bands with different mobilities would not explain the data, since R_H would be enhanced. This may explain the data for third-stage Cs-graphite in Figure 70.

Both the Hall coefficient and magnetoresistance [Q32] are consistent with a model in which electrons and holes are present. This
is not consistent with the model for the Fermi surface shown in
Figure 69, although a low density of holes is predicted by the model
where the cylindrical and spherical surfaces touch (Figure 69).
Further work is needed to clarify the reasons for the discrepancy.

The properties of very dilute compounds (both donors and
acceptors) have been treated on the basis of a rigid band model with
varying Fermi level, as discussed above for higher stage compounds.
The conductivity increase at low uptake has then been attributed to
the bulk of the sample whose carrier density has been increased by
the change in Fermi energy [Q19,Q23]. This is probably erroneous
[Q62]. A good case can be made for assigning the initial conductivity increase to graphite layer planes adjacent to intercalant
layers. These layers would have much higher conductivity than the
bulk of the layers removed from the intercalant. Another means of
expressing this is through the concept of charge donated per carbon
layer from the donor intercalant. Effectively, the donated charge
may reside in the adjacent layers, with relatively small value
donated to the planes further away. This is discussed further in
the next section.

B. Acceptor Compounds

The interpretation of the results for acceptor compounds is made more
difficult because band models have not yet been developed and galvanomagnetic coefficients are very difficult to measure because of the
extreme anistropy of the conductivity. The present comments are
therefore tentative, and conclusions are subject to change.

Early measurements of the conductivity of highly conducting acid
compounds were subject to considerable error because of the difficulty
of attaching leads to the samples. As a result of unequal sharing of
current between the planes, measured conductivities had to be lower
than in actuality. These problems have largely been overcome through

the use of an inductance technique (see Appendix A.4), although c-axis resistivity determination still remains problematical [Q84]. The resistivity of the sample might also depend sensitively on the perfection of the parent graphite (see Section XV.C).

At least in the early stages of intercalation the Hall coefficient is positive [Q7] indicating a lowering of the Fermi energy. But, Hall data are susceptible to gross inaccuracies when a four-point technique is used with a highly anisotropic metal, and data obtained in this way should be interpreted with caution. At the present time no satisfactory method of measuring the Hall coefficient has been devised for a contactless sample, although the Hall mobility may be obtained [Y12]. However, no measurements of the latter on acceptor compounds have been reported.

Data on the principal conductivities of several acceptor compounds of interest are presented in Table 4. Higher basal conductivities and anisotropies σ_a/σ_c are obtained for acceptor than for donor compounds. The highest basal conductivities are obtained for second- or third-stage compounds, rather than first-stage (see Figure 67 for $C_{8R}AsF_5$). Since anisotropy tends to reduce the measured value of σ over the intrinsic, it may be accepted that certain acceptor compounds have conductivity at least as high as that of copper, if not higher [Q81,Q84].

There has been considerable controversy over the magnitude of the fractional charge transfer, f, for these compounds. Ubbelohde [Q68] put forward the idea that, for graphite nitrate, one hole was generated for every four molecules of HNO_3 inserted, so that the resulting compound would have the formula $C_{6R}^{+}HNO_3^{-}\cdot 3HNO_3$. A possible density-of-states curve is indicated in Figure 71. This seemed reasonable in view of the large conductivity increase observed in this, and similar compounds. Consequently, f-values of this order of magnitude were generally assumed to hold for all acceptor compounds.

More recently, the fractional charge transfer for graphite bromine compounds has been evaluated as ~0.05 by M. S. Dresselhaus and G. Dresselhaus [Q18], for weak compounds (partly residue, partly

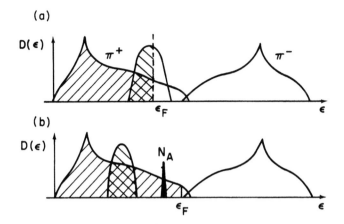

Figure 71.   (a) A hypothetical model for the density of states for
first stage graphite nitrate.   The acceptor level contains electrons
transferred from the π-bands.   The size of the neutral acceptor level
peak is magnified for clarity.   (b) Another model in which the intercalant levels are filled, but charge transfer occurs to a small density of acceptor levels, corresponding to an active species such as
HNO_3^-.

lamellar).   This implies that hole densities had been greatly over-
estimated and mobilities underestimated.   Other measurements have
indicated a larger value of f for higher stage compounds (~0.2-0.5).

    The discrepancy can be explained if it is assumed that the
magnetoreflection experiments on weak compounds were probing the
layers well removed from the layers of intercalant.   Thus, only a
small value of f was inferred.   However, the actual charge transfer
is higher, if allowance is made for the transfer to adjacent layers.
Recent experimental studies of the Fermi surface of $C_{8R}Br$, $C_{8R}AsF_5$,
and $C_{8R}SbF_5$ tend to confirm this model [Q3 and Q9].

    For the compounds of greatest current interest, namely C_nSbF_5,
C_nAsF_5, very little data are available except on the conductivity,
the magnetoresistance, and their dependence on temperature (Figure
72).   The decrease of resistivity with temperature of C_8AsF_5 and
$C_{16}AsF_5$ (first and second stages, respectively) is characteristic of
a metal.   Changes in slope of the $\rho(T)$ curves may be due to structural phase transitions, of the order-disorder type for example.

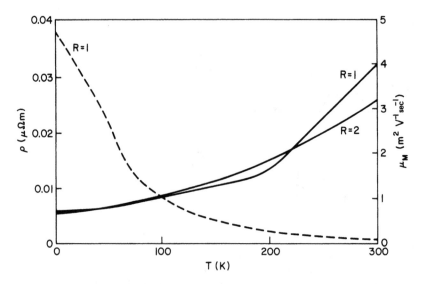

Figure 72. The variation of the electrical conductivity and magneto-resistance mobility (dashed line) with temperature for graphite inter-calated with AsF_5. (Data adapted from [Q85].)

However, the change in slope at ~10 K for $C_{16}AsF_5$ may be due to a change in the electron-phonon scattering process [A24], with $\rho \, \alpha T^5$ below and αT^2 above the transition temperature [Q85].

The magnetoresistance increases as B^2 at lower field values, but less rapidly at higher field ones [Q85]. The magnetoresistance even saturates at 4.2 K for the first-stage compound. The mobility, μ_M, is surprisingly high for these materials (Figure 72). If the density of carriers is estimated from $\sigma/e\mu_M = P$, then the fractional ionization is ~15-20%. This is in accord with recent NMR studies of the graphite-SbF_5 system by H. Resing and L. Vogel [ZQ17] which suggest that partial dissociation occurs:

$$3SbF_5 \rightarrow SbF_3 + 2SbF_6^- + 2 \text{ holes}$$

They estimated that $f \equiv P/SbF_5 \sim 0.2$. The corresponding density-of-states diagram is sketched in Figure 71b. The acceptor levels are associated with SbF_6^-, and the density, N_A, is equal to the density of this specie. Electronic levels associated with the other molecular species (SbF_5, SbF_3) are completely filled.

Based on the above experimental measurements the following guesses can be made about the electronic structure of $C_{16}AsF_5$:

1. The hole states are in the corner of the Brillouin zone, and the constant energy surfaces are nearly cylindrical, implying high effective masses in the c direction. This is consistent with the high conductivity anisotropy

2. The basal effective mass must be small ($\sim 0.1-0.2\ m_0$) to support high mobility values consistent with the magneto-resistance data.

3. In order for the relaxation time to be as high as possible, the density-of-states at the Fermi level $D(\varepsilon_F)$ must be low, and electron-phonon coupling constants small [see A25].

4. The density of holes $\sim 3.10^{26}\ m^{-3}$. This is roughly consistent with the recent evaluation of the size of the Fermi surface [Q3].

5. It is to be noted that a single-band model (hole band) would have very small magnetoresistance. Accordingly, there must be at least two groups of hole carriers, with different mobilities.

If the above conjectures are correct, the higher conductivity of acceptor compounds compared to donor is probably related to the smaller fractional charge transfer, and the consequent two-dimensional energy surfaces (cylindrical) for the acceptor compounds. Variability observed in the conductivity for a given stage of compound may be associated with different degrees of disproportionation. It is possible, therefore, that even higher conductivity may be possible through control of the fractional charge transfer.

C. Effect of Defects on Conductivity of Intercalation Compounds

The highly conducting, strong-acid compounds such as $C_{16}AsF_5$ have relatively high mobilities, of the order of $0.2\ m^2\ V^{-1}\ sec^{-1}$ at room temperature, when HOPG is the starting material. However, if practical synthetic metals are to be prepared, then a much less expensive

carbon, such as a fiber, will need to be used rather than HOPG. In
such a material the defect scattering becomes an important factor,
which may well limit the attainable conductivity to values well below
that of copper.

During the intercalation process the carbon layer planes must
"re-register" with respect to each other, a process controlled by
dislocation motion. Introduction of defects by fast neutron irradiation clearly hinders intercalation [Q61]. However, it is by no
means certain that defects are introduced by the uptake of intercalant, and in fact the reverse may be true. On the other hand,
Fischer [Q24] has speculated that the creation of defects by the
intercalation process may be increasingly severe as the strength of
the acid intercalant increases, and this may become a limiting factor
in the attainment of high conductivity.

If the data for resistivity and mobility (Figure 72) for $C_{16}AsF_5$
are analyzed using the graphite band model with Fermi level depressed
by acceptors, the mean free path for scattering at 4.2 K is estimated
to be ~3-6 µm. This is close to the value obtained for un-intercalated
HOPG of high quality (see Section IV). Accordingly, it is reasonable
to assume that scattering from crystallite boundaries controls the
mobility at low temperature. Using Mathiesson's rule (Equation A9
in Appendix A), it can be shown that the room temperature conductivity
of the same compound would be reduced by ~20-30% if poorer quality
HOPG were used (λ_d ~ 1 µm). This may help to partly explain the variability of room temperature conductivity observed for these compounds.

Fibers prepared from mesophase pitch precursor have been shown
to partially graphitize when heat-treated to ~3000°C (see Section IX).
Crystallite dimensions L_a ~0.03 µm have been measured. In this case
the room temperature conductivity would be reduced by approximately
a factor of 5 to 10. However, it is conceivable that L_a could be
increased by using catalysts for graphitization, and that fibers with
useful mechanical and electrical properties could be prepared. At
the present time the highest conductivity reported for an intercalated
fiber is far below that of copper [Q80], but recent results of L. Vogel
(unpublished) indicate that higher conductivities can be achieved.

APPENDIX A. PHENOMENOLOGICAL ASPECTS OF CONDUCTIVITY

The simple formula for the conductivity has already been introduced:

$$\sigma = Ne\mu \tag{A-1}$$

where σ is the conductivity, N the number of current carriers per unit volume carrying charge $-e$, and μ is the mobility of the charge carriers. The mobility can be simply envisaged as the mean drift velocity $< v >$ per unit electric field E applied to the carriers.

$$\mu = \frac{< v >}{E} \frac{m}{sec} \frac{m}{volts} = m^2 V^{-1} sec^{-1} \tag{A-2}$$

Although an equation of this sort is generally valid, it is to be recognized that different groups of carriers may have different mobilities and it is, therefore, better to express the conductivity as a sum over the contributions of these groups, j:

$$\sigma = \sum_j n_j e\mu_j + \sum_j p_j e\mu_j \tag{A-3}$$

For a semimetal such as graphite, a simple model supposes that there is a band of electrons (j = 1) and a band of holes (j = 2) (two-band model). The conductivity may then be written in terms of the total density of electrons, N, and holes, P:

$$\sigma = Ne\mu_n + Pe\mu_p \tag{A-4}$$

For the case of intrinsic graphite N is equal to P. Impurities and defects, however, may create acceptors or donors and decompensate the material. For instance, boron-doping and damage introduced by fast neutrons or electrons produces a predominance of holes ($P > N$).

1. Density of Carriers

The density of carriers can be determined from fundamental calculations of the energy bands. Such calculations give the density of
electronic states, $D(\varepsilon)$, (see Appendix C) from which the total number
of electrons, N, or holes, P, can be found ($N = \Sigma n_i$, $P = \Sigma p_i$)

$$N = \int_{\varepsilon_c}^{\infty} D_c(\varepsilon) f(\varepsilon) d\varepsilon \qquad\qquad (A-5)$$

$$P = \int_{-\infty}^{\varepsilon_v} D_v(\varepsilon) [1 - f(\varepsilon)] d\varepsilon \qquad\qquad (A-6)$$

where ε_c is the lowest energy in the conduction band, and ε_v the
highest in the valence band (Figure A-1). $f(\varepsilon)$ is the Fermi-Dirac
distribution function (see Ref. A25) which expresses the normalized
probability that an electron occupies a state. Thus, $[1 - f(\varepsilon)]$ in
Equation (A-6) represents the probability that the state is unoccupied.
This is equivalent to saying that a hole is present.

The above formulae hold generally, even for an imperfect material.
If donors and acceptors are present, the condition of charge neutrality
is expressed as

$$N + N_A^- = P + N_d^+ \qquad\qquad (A-6)$$

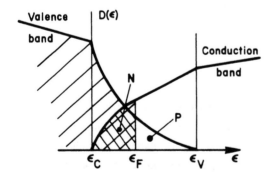

Figure A-1. The density of states for graphite: ε_c is the lowest
energy of the conduction band, ε_v the highest energy for the valence
band. The case shown is at T = 0K with N electrons and P holes.

where N_A^- is the density of acceptors with accepted charge and N_d^+ the
density of donors which have donated charge to the conduction band.
These formulae hold for both semimetals and semiconductors.

2. Mobility

It is very difficult to calculate the mobility from first principles.
However, it is usually possible to separate the intrinsic from the
defect part of the scattering. Simple theory shows that for any
group of carriers with mobility μ, and effective mass m^*, then

$$\mu = \frac{e\tau}{m^*} \qquad\qquad (A\text{-}7)$$

where τ is the relaxation time of the carriers, which is a measure of
the mean time between collisions. If the crystal is perfect, then
$\tau \equiv \tau_i$ is governed by the collisions of carriers with lattice vibrations (electron-phonon scattering). τ_i tends to very large values at
low temperature, because the amplitude of thermal vibrations tends to
a small value (the ground state vibrations at T = 0K are not effective
in scattering). (For a review, see Ref. A25.)

When defects are present, an extra scattering is produced (Figure
A-2), corresponding to a relaxation time τ_d. The total relaxation
time, τ_t, is given by Mathiesson's rule:

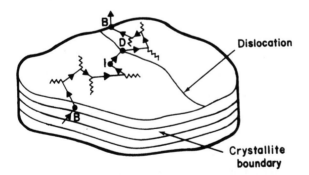

Figure A-2. Schematic of a path of an electron across a crystallite.
Scattering events are: ⋀⋀= phonons (intrinsic), B = boundaries,
D = dislocations, I = impurities.

$$\frac{1}{\tau_t} = \frac{1}{\tau_i} + \frac{1}{\tau_d} \tag{A-8}$$

Using Equation A-7, then mobilities sum as:

$$\frac{1}{\mu_t} = \frac{1}{\mu_i} + \frac{1}{\mu_d} \tag{A-9}$$

At low temperature, it is clear that $\mu_t = \mu_d$, since $\mu_i \to \infty$. Accordingly, an estimate of the mobility at low temperature can be useful for determining the density of defects present.

An important parameter characterizing the scattering of carriers is the mean free path between collisions, λ. If the mean carrier speed is $\bar{v}$, then

$$\lambda \sim \bar{v} \tag{A-10}$$

$$\mu = \frac{e\lambda}{m^*\bar{v}} \tag{A-11}$$

The concept of the mean free path is particularly important when considering defect scattering, since it roughly corresponds to the mean distance between the defects (mean free path due to defect scattering is denoted λ_d).

After a carrier has been scattered at a crystallite boundary, its direction of motion is statistically arbitrary. For graphite, the velocity component in the c direction, v_{11}, may be neglected compared to the basal component, $v\perp$, ($v_{11} < v\perp/100$). Accordingly, in the absence of any other type of scattering, the mean distance travelled before the carrier collides with another boundary is less than the diameter of the crystallite, L_a. For a circular crystallite $\lambda_d \sim L_a/2.5$, and a similar relationship is expected to hold for nearly circular, but irregular crystallites.

When the defect concentration becomes very high, λ_d becomes comparable to the interatomic dimensions. In this case the concept of

a free electron being scattered by defects becomes inappropriate.
Instead, the concept of a localized carrier is used, which is trans-
ported via a thermally assisted, or tunneling process. Localization
mechanisms, and the conductivity of defective semiconductor materials
are discussed in Appendix D.

3. Anisotropy of the Conductivity

Unlike a cubic metal, the conductivity of graphitic materials is
anisotropic. In fact, for single-crystal graphite the ratio of principal conductivities is higher than for any other elemental material.

The c axis of graphite is sixfold symmetric. Accordingly the
conductivity within the planes is independent of the direction of
current flow. The conductivity in the basal planes will be denoted
σ_a, and the other principal component in the c direction will be
denoted σ_c. Resistivities will be denoted ρ_a and ρ_c respectively.
The anisotropy ratio, r, is

$$r = \frac{\rho_c}{\rho_o} = \frac{\sigma_a}{\sigma_c} \tag{A-12}$$

If current flows with density j_a in the planes, and j_c in the
c direction, the total current density is $\sqrt{(j_a^2 + j_c^2)^{-2}}$. The electric
field components developed across the sample will be:

$$E_a = \rho_a j_a \tag{A-13}$$

$$E_c = \rho_c j_c \tag{A-14}$$

The resultant electric field vector $\overline{E}$ with components E_a and E_c
is not parallel to the current vector $\overline{J}$ with components j_a and j_c as
illustrated in Figure A-3. Where practical, experiments are carried
out with current flow in principal directions. This sometimes creates
difficulties, as outlined in the next section.

The anisotropy ratio for single crystal graphite is usually of
the order of $r \sim 100$. As discussed in the text, values as high as

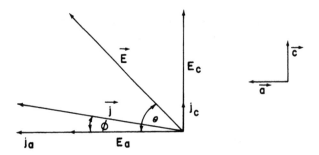

Figure A-3. Vector representation of the electric field in an anisotropic conductor at angle θ to the a-axis, and the resulting current
vector at angle ϕ. If $\theta > \phi$, then $\sigma_c < \sigma_a$. ($\sigma_c = j_c/E_c$; $\sigma_a = j_a/E_a$.)

$r \sim 10^5$ have been found in highly oriented pyrolytic graphite at low
temperatures. Even higher values have been found in intercalated
samples (see Section XV). The anisotropy ratio is strongly dependent
on the defects present in the material. The degree of preferred
orientation of crystallites in a polycrystalline material can change
the anisotropy ratio by orders of magnitude. If crystals are oriented
at random, the bulk sample may have isotropic resistivity characteristics as in heat-treated carbons. At the microscopic level, however,
the conductivity within each crystallite will be anisotropic, so that
the current flows along a highly erratic path of lowest resistance.
Further discussion is given in Section XIII.

When a sample is placed in a magnetic field the current and electric field components are more complicated than for the zero magnetic
field case discussed above. A discussion is given in Appendix B.

4. Measurement Techniques

It is very difficult to make reliable measurements of the conductivity
and galvanomagnetic properties of graphite for a number of reasons.
Firstly, high quality graphite crystals are very fragile, so that
great care must be taken in attaching leads [F9]. Since the conductivity of graphite is very high (2.5×10^{-6} Ωm^{-1} at 298 K), a *four-
point* or comparable technique must be used to avoid contact resistance
effects (Figure A-4). A third difficulty is associated with the
anisotropic conductivity. As a result, it is very difficult to

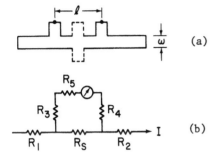

Figure A-4. (a) Geometry of a sample for four-point resistivity measurements. Dotted lines are additional arms for Hall measurements. (b) Equivalent resistance network. R_s is sample resistance to be measured. $R_3 + R_4 \ll R_5$, and $R_5 \gg R_s$, where R_5 is the internal resistance of the voltmeter.

ensure that current is equally shared by all the planes when basal properties are being measured.

For most basal measurements a specimen with side arms is preferred (Figure A-4). Both current and potential leads must be attached in such a way that contact is made to all the planes. Usually this is accomplished by soldering onto copper-plated tips, or with the use of silver-epoxy adhesive.

The importance of equal sharing of current between the planes can be illustrated with a useful transformation to an equivalent isotropic solid. Consider a cube of material with conductivity anisotropy r. A block of material with isotropic conductivity σ_a would have the same electrical characteristics if the dimension of the high resistivity direction were multiplied by $r^{1/2}$.[*] As a rule of thumb, the length-to-thickness ratio of an isotropic conductor for conductivity measurements must be greater than about 5 to avoid possible end effects (see Figure A-4). As seen in Figure A-5, the cross-section of a specimen cut with length-to-thickness ratio of 10 appears as a specimen with square cross-section in its equivalent

---

[*]This transformation gives the correct resistance ratio in basal to c axis directions. In order to obtain the correct sample resistance, the basal dimension should be reduced by the factor $r^{1/2}$.

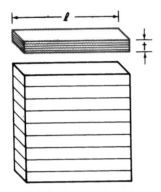

Figure A-5. The anisotropic specimen with length-thickness ratio,
ℓ/t, = 10, and anisotropy ratio r = 100. The equivalent isotropic
solid (r = 1) is shown below ($\ell/t \rightarrow 1$).

isotropic form if r = 100. This implies that great care must be
taken to apply current leads, so that they contact all of the end
area.

The effect of applying current to specific points on the ends
of the sample can be explored using conducting paper and simple
voltage measuring equipment. Some current flow lines and equipotential lines are indicated for a particularly severe case in Figure
A-6. Whereas the experimenter requires the current flow to be
parallel to the planes, in fact it can be nearly perpendicular to
this in the interior of the specimen, as illustrated. In the actual,
anisotropic, graphite sample the current apparently flows in thin
layers near the basal surfaces before passing between these surfaces,
in order to minimize Joule heating. Obviously, the measured conductivity will depend very sensitively on how nearly the experimenter
achieves the desired current flow.

In general, geometrical effects such as those discussed above
will result in a measured value of ρ_a which is higher than the
actual value, since $\rho_c \gg \rho_a$.

When small samples of graphite are being used a major source of
error in the determination of ρ_a is associated with the effective
length-to-width ratio (ℓ/w in Figure A-5). If larger samples are

(a)

(b)

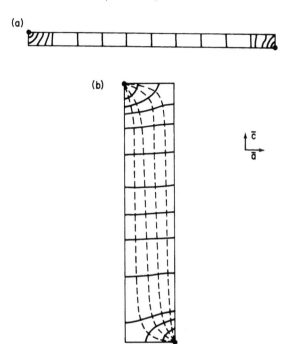

Figure A-6. (a) A sample with isotropic conductivity and $\ell/w = 20$ in which current contacts are at opposite corners. This has negligible effect on the equipotential lines in the interior. (b) The equivalent isotropic sample if the anisotropy were 10,000. The basal dimension of the sample is reduced x5 for convenience. In this case the current lines (dashed) are nearly at right angles to the desired flow direction in the interior. (Figure adapted from Spain, Ph.D. Thesis, Univ. of London, 1964.)

available, an accurate determination of ρ_a can be made using the Van der Paauw technique [Y13]. This only requires the sample to be of uniform thickness. The shape of the specimen in the plane of the current flow is not important. However, the leads should be attached with great care to ensure homogeneous sharing of current between the planes. The experimenter is also advised to determine the optimum position for the leads with a dummy sample of conducting paper of roughly the same shape as the real sample, since the error of the measurement can be minimized with a judicious choice of contact positions (see Ref. F2 for application to HOPG).

An extension of this technique, called the Montgomery technique [Y9], in principle can be used to give both ρ_a and ρ_c. However, it does not work well with a material such as crystalline graphite in which the anisotropy ratio is so high.

It is interesting to note that the anisotropy ratio r can become extremely high when samples are intercalated (e.g. $\gtrsim 10^6$). In this case it is very difficult to attach the current leads in such a way that current is shared equally between the planes. An r.f. induction technique can be used satisfactorily in this case [Q84]. It is difficult to calculate the conductivity from first principles using this method, but it can give reliable comparisons of conductivity between the sample and a standard sample (e.g., copper), chosen with conductivity approximately equal to ρ_a of the sample to be measured. This technique, and a comparison of other techniques for these highly anisotropic materials is given in Refs. Q84, Y14, and ZY1.

For c-axis current flow, the anisotropy makes the experiments much simpler. The main limitation is in the size of available samples, particularly the c dimension. Also, it is impractical to construct a bridge-shaped specimen, such as that used for basal conduction. The potential experimenter is cautioned not to use the same contacts for current and potential leads. Such a two-point measurement invariably gives a grossly enhanced estimate of the resistivity because of the inclusion of contact resistance.

APPENDIX B. GALVANOMAGNETIC EFFECTS

The definitions of the Hall resistivity, $\rho_{yx}(B_z)$, and the magneto-resistivity, $\rho_{xx}(B_z)$ are illustrated in Figure A-7. The Hall coefficient, R_H, is equal to ρ_{yx}/B_z. The magnetoresistance can be measured with the magnetic field in other directions, to be discussed at the end of this Appendix. These quantities are conveniently defined for a rectangular specimen, but other geometries can be used.

The relationships between ρ_{yx} and ρ_{xx} and the carrier densities and mobilities can be found for simple models of the dispersion

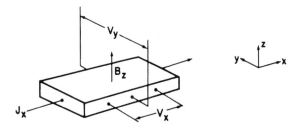

Figure A-7. Definition of the Hall resistivity, ρ_{yx}, Hall coefficient, R_H, and magnetoresistivity, ρ_{xx}; where $\rho_{xx} = E_x/j_x = (V_x/J_x)$
$(S_y S_z/S_x)$, $\rho_{yx} = E_y/j_x = V_y S_z/J_x$, and $R_H = \rho_{yx}/B_z = V_y S_z/J_x B_z$.

relationship and relaxation time. Two simple models will be considered here.

1. Simple Two-Band Model

The simple two-band model (STBM) assumes that there are bands of
electrons and of holes, both described by parabolic energy-wavevector
relationships. The most important equations are

$$\frac{\Delta\rho_{xx}(B_z)}{\rho_{xx}(0)} = \frac{[(1 + b)NPb/(N + bP)^2]\mu_n^2 B^2}{(N - P)/(Nb + P)^2 b^2 \mu_n^2 B^2 + 1} \qquad (A\text{-}15)$$

$$R_H(B_z = 0) = \frac{-1}{e}\frac{(N - Pb^2)}{(N + Pb)^2} \qquad (A\text{-}16)$$

$$R_H(B_z \to \infty) = \frac{-1}{e(N - P)} \qquad (A\text{-}17)$$

where

$$b \equiv \mu_p/\mu_n$$

A sketch of the variation of ρ_{xx}, ρ_{yx}, R_H is given in Figure A-8,
for the particular case $N = 1.02\,P$, $\mu_p = 1.1\,\mu$. Several points are
to be noted.

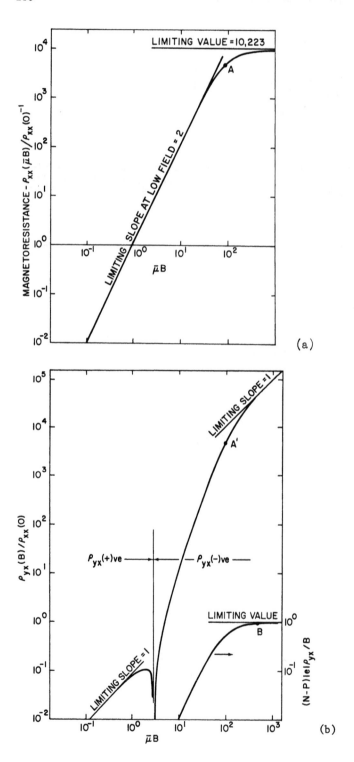

(a)

(b)

a. At low field, the denominator in (A-15) does not deviate appreciably from unity if

$$\left(\frac{N - P}{Nb + P}\right)^2 b^2 \mu_n^2 B^2 \gg 1$$

Thus the magnetoresistance should be proportional to B^2. In this region, the geometrical mean mobility $(\mu_n \mu_p)^{1/2} \equiv b^{1/2} \mu_n \equiv \mu_M$ can be obtained from

$$(\mu_n \mu_p)^{1/2} = \left[\frac{(N + bP)^2}{(1 + b)NP} \frac{\Delta\rho}{\rho_0 B^2}\right] \qquad (A-18)$$

If the ratios $b = \mu_n/\mu_p$ and N/P lie reasonably close to unity then

$$\mu_M \equiv \left(\frac{\Delta\rho}{\rho_0 B^2}\right)^{1/2} \sim (\mu_n \mu_p)^{1/2} \qquad (A-19)$$

This equation has been widely used to estimate the mobility in graphite.

b. At sufficiently high field the magnitude of $|\rho_{yx}|$ becomes greater than ρ_{xx}. At higher fields ρ_{xx} tends to a saturation value

$$\frac{\Delta\rho}{\rho_0}(\text{sat.}) = \frac{(1 + b) \ NP \ (Nb + P)^2}{(N - P)^2 \ b \ (N + bP)^2} \qquad (A-20)$$

provided

$$\left(\frac{N - P}{Nb + P}\right)^2 b^2 \mu_n^2 B^2 \gg 1$$

Figure A-8. A plot of $\rho_{xx}(B)$ and $\rho_{yx}(B)$ in reduced coordinates. (a) $\rho_{xx}(B)/\rho_{xx}(0)$ vs. $\bar{\mu}B$ $(\bar{\mu} = \sqrt{\mu_n\mu_p})$. At point A, $|\rho_{yx}| = \rho_{xx}$ and ρ_{xx} saturates at higher field; (b) $\rho_{yx}/\rho_{xx}(0)$ and $\rho_{yx}|e|(N - P)/B \equiv R_H(B)/R_H(B = \infty)$. At point A´ $|\rho_{xy}| = \rho_{xx}$ and ρ_{yx} then tends to its limiting slope while the Hall coefficient saturates. Curves are drawn for the two-band model with $N = 1.02 \ P$, $\mu_p = 1.1 \ \mu_n$.

Also, R_H tends to the saturation value given in Equation A-17. This gives in principle a valuable method of determining the difference in carrier densities $(N - P)$. For perfect graphite $N - P = 0$, so that nonzero values are related to the presence of defects.

c. If the Fermi level is raised or lowered by adding defects, such as impurities, the variation of the Hall coefficient at low field can be followed easily using Equation A-16. A sketch is given in Figures 49, 51, 53. As $|P - N|$ is increased, the value of R_H initially falls for donor impurities, rises for acceptor impurities. At sufficiently high doping levels, the curves of R_H as a function of $(P - N)$ approach the single-carrier limit equations to be discussed next. Thus $R_H(P - N)$ curves have maxima and minima.

d. For the two-band model the Hall mobility can still be defined

$$\mu_H \equiv R_H/\rho_{xx}(0) \equiv \sigma R_H \tag{A-21}$$

However, μ_H will not represent a physically useful quantity. The inequality $\mu_H \gg \mu_M$ which pertains for the two-band model can be used as a criterion to indicate whether a two-band or single-band model is applicable.

e. The two-band model was applied above to bands of electrons and holes. It can also be used for the case of two bands of electrons or holes with different properties; e.g., a group of highly mobile and another of less mobile electrons. The relevant equations for the Hall effect or for magnetoresistance are simple adaptations of those given for hole and electron bands.

2. Single-Carrier Model (Single-Band Model)

If the Fermi level is depressed or raised beyond the band edges, the density of one type of carrier with respect to the other will become sufficiently small so that only one carrier needs to be considered. Then, simple theory shows that this for all values of B, at T = OK.

$$\left(\frac{\Delta\rho}{\rho_0}\right) = 0 \qquad \text{i.e. } \mu_M = 0 \qquad\qquad\qquad \text{(A-22)}$$

$$R_H = \begin{cases} \dfrac{-1}{Ne} & \text{for electrons} \\[2mm] \dfrac{+1}{Pe} & \text{for holes} \end{cases} \qquad\qquad \text{(A-23)}$$

In this case, the Hall mobility gives a reasonable representation of
the drift mobility

$$\mu_H = \frac{R_H}{\rho_{xx}(0)} = \begin{cases} \mu_n & \text{for electrons} \\[2mm] \mu_p & \text{for holes} \end{cases} \qquad\qquad \text{(A-24)}$$

At higher temperature $(T > 0)$ weak magnetoresistance is predicted
by this model. However, the condition $\mu_M \ll \mu_H$, can be used as a
criterion for judging whether the single-band model is applicable.
Another contribution to the magnetoresistance can come about from
the warping of the constant energy surfaces [D3]. However, such contributions are relatively weak and the inequality $\mu_M \ll \mu_H$ still holds
if there is only one band of carriers.

Neither of the simple models discussed above are applicable to
single-crystal graphite. This is because the dispersion relationship
is extremely complicated. Although the STBM can be used to roughly
estimate carrier mobilities and densities, there are serious difficulties with the interpretation of galvanomagnetic data [F2]. Trends
in the magnetoresistance can be interpreted satisfactorily for different specimens in terms of defects, however.

3. Magnetoresistance Anisotropy

As mentioned earlier, the magnetoresistance depends on the direction
of the magnetic field with respect to the electric current and crystal
areas. The principal effects are illustrated in Figure A-9 for single-
crystal graphite. When the magnetic field is in the longitudinal or
transverse position parallel to the planes, the magnetoresistance is

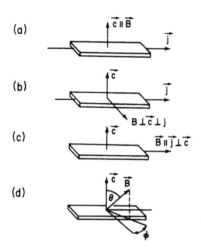

Figure A-9. The principal magnetoresistance coefficients of single-
crystal graphite, or HOPG: (a) Principal transverse effect;
(b) secondary transverse effect; (c) longitudinal effect; (d) general
case--usually the angle θ only needs to be specified since longitudinal and secondary transverse effects are dominated by misorientation
effects.

very small compared to the "normal" effect with $\overline{B}\|\overline{c}$. Rotation
diagrams are illustrated in Figures 18, 29 and 45. A fit to the
data cannot be obtained with a sum of $\sin^2\theta$ and $\cos^2\theta$ terms. A
satisfactory explanation for the angular dependence has not been
obtained, due to the complicated dispersion relationship.

The main importance of the rotation diagram is for a simple
assessment of preferred orientation. Consider a sample composed of
many small crystallites with c axes misoriented with respect to the
net c direction (Figure 9). If the magnetic field lies perpendicular
to the net c direction, individual crystallites will experience a
small component of magnetic field parallel to the c axes. As a
result, the bulk magnetoresistance will correspond approximately to
the normal magnetoresistance ρ_{xx}(B < cosδ >), where B < cosδ >
represents the average component of field parallel to individual
c axes. Thus, the measurement of either the longitudinal, or the
transverse magnetoresistance with magnetic field perpendicular to

the net c axis, will give a measure of the degree of preferred
orientation. This technique is of use for pyrolytic graphites but
is limited for less perfect specimens because of the onset of negative magnetoresistance.

4. Negative Magnetoresistance

Negative magnetoresistance has been considered in detail by Delhaes
in Volume 7 of this series [D2]. A brief summary will be given here
with discussion of a model that has recently been applied to under-
stand negative magnetoresistance in carbon fibers.

Considering a crystalline material, negative magnetoresistance
can only occur if (see Equation 1):

1. Magnetic field increases the density of carriers
2. Scattering rates are reduced by an applied magnetic field,
   so that the mobility is increased.

A further explanation for negative magnetoresistance is possible
for an inhomogeneous material. This inhomogenity can be of the form
of spatial variations in the density of carriers, for example.
Another possibility is that internal "size" effects at boundaries
between crystallites produce diffuse scattering of carriers, resulting in negative magnetoresistance [H6].

An increase in mobility with magnetic field could result from
the presence of localized spins in the material, which align in the
magnetic field [A20]. An alternative model suggests that the Fermi
energy may be shifted by magnetic field in the vicinity of the
mobility-gap edge (see Figure A-19 in the Appendixes, with ε_F near
ε_v, for example), producing an enhancement of mobility.

Delhaes et al. [G7] carried out a detailed examination of the
possible origins of magnetoresistance for heat-treated pyrocarbons;
some results are presented in Figures 24 and 25. They were not able
to separate the positive and negative components of magnetoresistance,
but found that the magnetoresistance could be scaled over certain
magnetic field ranges and that negative magnetoresistance was anisotropic. Their results favored a model based on a mobility gap rather

than scattering from localized spins. However, none of their models were able to explain the results satisfactorily.

Yazawa [G32] proposed a model for the negative resistance based on the increase in carrier density with magnetic field. He argued that for turbostratic graphite the electronic structure should be nearly that of 2D graphite. In a magnetic field the Landau level structure (see Appendix C) is very different from that in 3D graphite. A large separation between Landau levels occurs at relatively low field [G29-G32], so that the quantum-limit condition can occur at only modest field values. In this condition, all the free carriers lie in the ground state magnetic level whose degeneracy (the density of electronic states in each magnetic level) increases with B, so that the carrier density can increase.

The increase in the density of carriers taking part in the conduction process is not limited to 2D graphite. The condensation of electronic states into magnetic energy levels for 3D graphite gives rise to the Shubnikov-de Haas oscillations (see Appendix C) and in fields above ~2T the carrier density increases with magnetic field. Above ~7T this increase is linear in field, but the magnetoresistance is strongly positive in this field range because the increase in carrier density is outweighted by the normal increase in resistivity. It is the very low value of magnetic field required to bring about this condition in 2D graphite that is remarkable.

The main features of this model, as developed by Bright [K4, ZD1], are presented in Figure A-10. In zero magnetic field a small overlap of conduction and valence bands was assumed, with Fermi level depressed due to the acceptors at $\varepsilon = \varepsilon_a$. An extra density of states N_0 was postulated to model the effects of partial 3D ordering (dotted line in Figure A-10b).

In a magnetic field, the density of states in perfect 2D graphite would be represented by the δ-functions (Landau levels) in Figure A-10c. As B increases, the spacing between the levels increases (see Equation A-37 in Appendix C). Due to defects the δ-functions

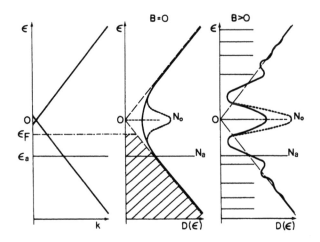

Figure A-10. Schematic of the model developed by Bright [K4] from Yazawa [G32] to explain negative magnetoresistance. (a) $\varepsilon(k)$ for 2D model; (b) density of states (B = 0); (c) density of states, B > 0. (Figure from [K4].)

are broadened. Using the Heisenberg uncertainty principle we expect the broadening, $\Delta\varepsilon$, to be roughly $\Delta\varepsilon\tau \sim h$, where τ is the relaxation time for carrier scattering. Bright then calculated the magneto-resistance to fit experimental curves for the magnetoresistance of fibers by optimizing values of parameters in his model, such as N_a, N_0. The agreement is remarkable (Figure A-11).

Bright's development of Yazawa's model is certainly too simplistic to be correct, yet it represents an important advance in our quantitative understanding of the phenomenon of negative magneto-resistance. Further work is required to assess the importance of sample inhomogeneity on a microscopic scale (e.g. regions of ~2D and ~3D material), the role of localized states and bands of defect states. Also, it will be interesting to see whether the model can account for the negative magnetoresistance in other carbons and phenomena such as the increase of negative magnetoresistance of irradiated samples of graphitizable carbons with annealing [M2].

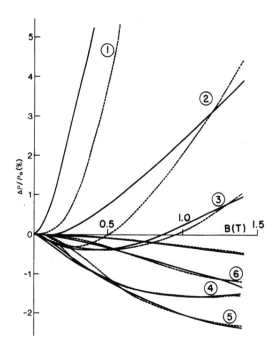

Figure A-11. Comparison of experimental and calculated values of magnetoresistance for fibers prepared from mesophase pitch precursor. (Figure adapted from [K4].)

5. *Kohler's Rule and Scaling Concepts*

The magnetoresistance of many metals can be scaled onto a universal curve using Kohler's rule [A25].

$$\frac{\Delta\rho}{\rho_0}(B,T) = F(B/\rho_0(T)) \qquad (A-25)$$

where ρ_0 is the zero value of ρ_{xx}. Since the density of carriers varies appreciably with temperature for graphite and carbons compared with a metal, where it is usually constant, a more suitable scaling relationship would be [D16]

$$\frac{\Delta\rho}{\rho_0}(B,T) = f(\overline{\mu}(T)B) \qquad (A-26)$$

where $\overline{\mu}$ is the mean mobility of the carriers.

The STBM expression for the magnetoresistance (Equation A-15) does not scale according to Equation A-25, unless P = N. In this case, the simplified Equation A-19 holds, and $\overline{\mu}$ is the geometric mean mobility $(\mu_n \mu_p)^{1/2}$. Experimental data show clearly for NSCG [E15] and HOPG [D26] that Kohler's rule does not hold in its modified form [A-26], although a discussion of data for KG has been given in terms of this rule [D16].

Undoubtedly, the complicated nature of the dispersion relationship and scattering processes for the free carriers ensures that the magnetoresistance has a complicated dependence on temperature and magnetic field. It is therefore not surprising that Kohler's rule is not applicable. However, the use of scaling concepts is useful, particularly the comparison of data at different temperatures, plotted as a function of $\mu_M^* B$ for example. As shown in Equation 8, μ_M^* is defined using scaling concepts. Unfortunately, the value of mobility to be used in these scaled plots cannot be defined uniquely, as discussed in Section IV, and in more detail in Refs. D26 and F2.

Scaling concepts have been used in Figures 38 and A-8, and have been applied to KG [D16], HOPG [K8], pyrolytic carbons [G7] and glassy carbons [J10].

APPENDIX C. DENSITY OF STATES AND DISPERSION RELATIONSHIP OF THE CARRIERS

The dispersion relationship of the carriers $\varepsilon(k)$ describes the variation of the electron energy ε with wavevector $k = 2\pi/\lambda$, where λ is the wavelength of the electron. When this relationship is known the density of states $D(\varepsilon)$ and other dynamic properties such as the electron velocity, can be determined. In conjunction with models for the variation of the carrier relaxation time, $\tau(k)$, the conductivity and galvanomagnetic properties can then be obtained using standard techniques [A25]. The dispersion relationship is applicable to a crystalline material. For a disordered material the concept of density of states is still valuable, but the wavevector k may not be a useful quantity (see Equation 2).

The dispersion relationship has been calculated for graphite using different techniques (see References, Section C, and review in Ref. A18). In these calculations use is made of the concept of a Brillouin zone, illustrated in Figure A-12 for graphite. The faces of the Brillouin zone represent the values of k for which the electrons are diffracted by the crystal. Within the zone, the allowed values of k lie on a grid, with one allowed state of given spin direction in volume $(8\pi^3)^{-1}$ in k-space, for unit volume of crystal. Accordingly, the Brillouin zone of graphite has sufficient volume to accomodate one electron for two atoms, for both spin directions.

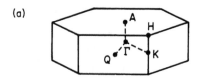

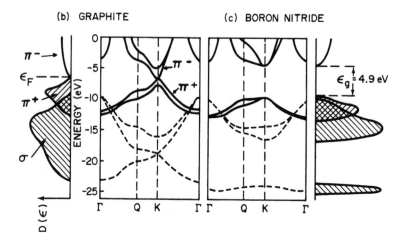

Figure A-12. (a) Brillouin zone of graphite showing symmetry points; (b) energy variation along symmetry lines ΓQ, QK, KΓ for graphite, and a sketch of the density of states. Solid lines are for π-electrons, dashed lines for σ-electrons. Zero energy is that of an electron with zero kinetic energy in free space; (c) the same diagram for hexagonal BN showing the large band gap. (Figures (b) and (c) adapted from [C36].)

Calculations of $\epsilon(k_x, k_y, k_z)$ are usually made along high symmetry directions, such as ΓK, ΓQ, etc., and results are presented in Figure A-12 for both graphite and hexagonal boron nitride. Note that both σ- and π-electrons are included (four electrons per atom). Since there is one π electron per atom, there are two branches of $\epsilon(k)$ for the π electrons in the Brillouin zone, while the σ-electron branches are doubly degenerate. The separation of the two π-electron branches is due to interactions between the layer planes.

For graphite, one branch of the dispersion curve for π-bonding states just touches the π antibonding branch at the symmetry point K. The electron energy varies along the edge of the Brillouin zone (KH) by about 40×10^{-3} eV, and this causes the valence band (π-bonding) to overlap the conduction band (π-antibonding) by this energy. Accordingly, graphite is a semimetal with free electrons and holes at all temperatures. By contrast, hexagonal BN is a semiconductor with a large band gap (~4.9 eV).

A useful concept is that of the surfaces joining states of the same energy in the Brillouin zone (equienergy surfaces). These are sketched for one branch of the valence band in Figure A-13. The surfaces are nearly cylindrical over a large portion of the zone, reflecting the anisotropic nature of the crystal. (An isotropic crystal would have spherical constant energy surfaces.)

Conduction in graphite can be related to the electrons and holes near the overlap region. At T = 0K the electrons fill all states up

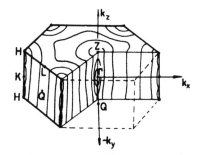

Figure A-13. The Brillouin zone of graphite showing symmetry points. Some constant energy surfaces are projected onto three planes. (Figure from [Q62].)

to the Fermi energy, ε_F, and the equienergy surface for ε_F (the Fermi surface) is sketched in Figure A-14. It is highly elongated and trigonally warped.

The carrier velocity at any point in the Brillouin zone can be calculated from the relationship

$$\bar{v} = \frac{1}{\hbar} \frac{\overline{\partial \varepsilon}}{\partial k} \tag{A-27}$$

Since the dispersion relationship for electrons in a crystal is different from that of free electrons ($\varepsilon = \hbar^2 k^2 / 2m_0$, where $m_0 = 9.11 \times 10^{-31}$ kg), the carrier speed for a given value of k is different also ($v = \hbar k / m_0$ for the free electron). A useful parameter for expressing this difference is the effective mass, m^*, for the electrons in a crystal. The definition of m^* for the conductivity is

$$m^* = \hbar^2 \, k \, \frac{dk}{d\varepsilon} \tag{A-28}$$

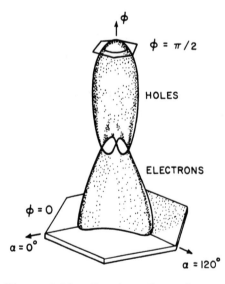

Figure A-14. Fermi surfaces for graphite, located along the vertical edge of the Brillouin zone (KH in Fig. A-12). The dimension in the plane is magnified x 5 for clarity.

The effective mass will depend on the direction in which the deriva-
tive $dk/d\varepsilon$ is taken. In principal directions, for electrons near
the Fermi energy, m^* (basal) varies between ~0.004-0.06 m_0 along the
zone edge, so that carriers are very light compared to the free electron case, whereas m^* (c direction) ~5 m_0. Using formula (A-7) for
the mobility, this suggests a mobility anisotropy of at least 100
for graphite even with an isotropic relaxation time.

Considering a general dispersion relationship such as those
indicated in Figure A-12 for graphite and hexagonal BN, the density
of states (the number of allowed energy states per unit energy interval) can be obtained from

$$D(\varepsilon) = \frac{dN(\varepsilon)}{d\varepsilon} \tag{A-29}$$

$$N(\varepsilon) = \frac{1}{4\pi^3} \int dk_x \, dk_y \, dk_z \tag{A-30}$$

where $N(\varepsilon)$ is the total number of states per unit volume of crystal
with energy less than ε. The integral over k-space uses the fact
that in volume $d^3 k$, there are $d \, k^3/4 \, \pi^3$ states of both spin directions. The integration is taken over states with energy less than
ε, and represents the volume enclosed within the equi-energy surface
(see Figure A-13). The volume can be evaluated if $\varepsilon(k_x, k_y, k_z)$ is
known.

The properties of electrons near the band overlap region are
particularly important for conduction processes. The density of
states for crystalline 3D graphite in this region has been discussed
in detail in [A18]. The dispersion relationship is complicated, and
$D(\varepsilon)$ unusual (Figure A-15a). Note the change in the form of $D(\varepsilon)$ for
the conduction and valence bands within and outside the region of band
overlap, as a result of the truncation of the constant energy surfaces
by the upper and lower surfaces of the Brillouin zone ($k_z = \pm\pi/c$).
Note also the concave-downwards shape of the density of states for
the valence band within the overlap region, which is unusual.

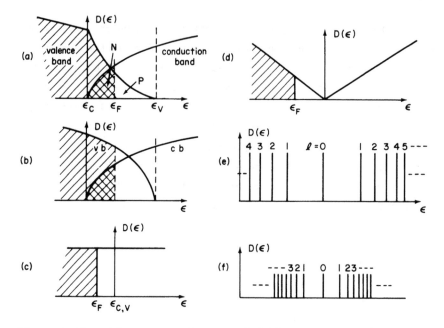

Figure A-15. (a) Density of states [D(ϵ)] for graphite (Slonczewski-Weiss-McClure model); (b) D(ϵ) for ellipsoidal model; (c) D(ϵ) for cylindrical, parabolic $\epsilon(k)$ model; (d) D(ϵ) for cylindrical, linear $\epsilon(k)$ model (Haering-Wallace model); (e) D(ϵ) for Haering-Wallace model B > 0; (f) same as (e) but higher value of B.

If the dispersion relationship for holes and electrons were ellipsoidal, e.g., for electrons

$$\epsilon = \frac{\hbar^2}{2m_\perp}(k_x^2 + k_y^2) + \frac{\hbar^2}{2m_{||}} k_z^2 \tag{A-31}$$

where $m_\perp$ and $m_{||}$ are effective masses perpendicular and parallel to c axes respectively, then

$$D_c(\epsilon) = \frac{1}{2\pi^2} \frac{(2m_\perp^2 m_{||})^{1/2}}{\hbar^3} \epsilon^{1/2} \tag{A-32}$$

similar to the curve for the electrons in Figure A-15a in the region
of band overlap. If the constant energy surfaces are approximated
by cylinders for energies outside the band overlap region with

$$\varepsilon = \frac{\hbar^2}{2m_\perp}(k_x^2 + k_y^2) = \frac{\hbar^2}{2m_\perp}\kappa^2 \qquad\qquad (A\text{-}33)$$

then

$$D(\varepsilon) = \frac{2m_\perp}{\pi c \hbar^2} \qquad\qquad (A\text{-}34)$$

which is independent of energy. This is roughly the case in Figure
A-15a for energies outside the band overlap region.

For 2D graphite the constant energy surfaces are cylinders, but
the dispersion relationship is different from (A-33) [C8,C32] (Haering-
Wallace model).

$$\varepsilon = \frac{\sqrt{3}}{2} a \, \gamma_0 \kappa \qquad\qquad (A\text{-}35)$$

where γ_0 is the in-plane overlap integral (~3 eV). The corresponding
density of states is:

$$D(\varepsilon) = \frac{8\varepsilon}{3\pi a^2 \gamma_0^2 c} \qquad\qquad (A\text{-}36)$$

which is linear in ε (Figure A-15d). At higher conduction band, or
lower valence band, energies the density of states for 3D graphite
asymptotes to that of 2D.

For either the 3D or 2D band structure, the density of carriers
at room temperature would not change by more than about a factor of
two if ε_F were shifted by donor or acceptor impurities, for $\Delta\varepsilon_F \lesssim$
±0.05 eV. This is because thermal excitation of carriers smooths
out the changes in carrier densities until ε_F is well removed from

the band overlap (3D) or band contact (2D) region. This implies
that relatively heavy doping levels ($\gtrsim$1000 ppm) are required to
increase carrier density appreciably.

One other interesting point about $D(\varepsilon)$ for 3D graphite is the
relatively higher value for the valence band than the conduction
band, outside the band overlap region (Figure A-15a). (This is
mainly due to the parameter γ_4 in the SWMcC dispersion relationship
[C28].) Higher values of $D(\varepsilon)$ correspond to higher effective mass
values (see Equations A-30 or A-33). Accordingly, for an intercalation compound in which ε_F is shifted upwards (donor compound) or
downwards (acceptor) by the same value, the effective mass and density of states will be higher for the acceptor compound. Since the
relaxation time for electron-phonon scattering is proportional to
the inverse of the density-of-states (see Refs. A18 and A25), donor
compound should have a lower effective mass, longer relaxation time,
and higher conductivity than the acceptor compound. This is the
reverse of what is observed (see Section XV).

In a magnetic field the energy states of the electrons coalesce
into magnetic energy levels or "Landau" levels [A25]. For 2D graphite
the energy levels are given by a simple formula:

$$\varepsilon_m = \pm \frac{\sqrt{3}}{2} \gamma_0 \left(\frac{eB}{\hbar}\right)^{1/2} \sqrt{\ell} \qquad\qquad (A-37)$$

where ℓ is the index for the level ($m = 0,1,2,3...$). These levels
are shown for two values of magnetic field in Figure A-15d, e. For
a normal 3D material the corresponding formula would be

$$\varepsilon_m = (\ell + 1/2) \frac{eB}{m^*} \qquad\qquad (A-38)$$

which is linear in ℓ. It is the square root dependence for 2D
graphite which produces the large separation of the Landau levels
at low values of ℓ. If $\gamma_0 = 3.2$ eV, $a = 2.46$ Å then the spacing
between the $\ell = 0,1$ levels is 37.5 meV for $B = 1$ T. If the 3D

formula (A-38) were used, an effective mass $m^* = 2.95 \times 10^{-3} m_0$ would be required to give the same level spacing at B = 1 T. Thus, the 2D structure can give effects very like those of low-mass carriers when a magnetic field is applied [G29].

APPENDIX D. CONDUCTION PROCESSES IN DISORDERED MATERIALS

Conduction processes in disordered materials are very different from those in crystalline, since the carriers can be localized by disorder. Most models have been developed for other materials—principally amorphous Ge and Si, and chalcogenide glasses. However, the ideas contained in these models can at least be applied qualitatively to carbon. As Mrozowski has pointed out [H17], it is astonishing that modern texts which review the properties of disordered materials or the physics of the phenomena (see for instance Refs. A13, A14) do not discuss the material most important from an industrial standpoint—carbon!

Consider the simplest possible case of a crystalline semi-conductor, such as Ge or Si, with valence and conduction band edges separated by a band gap ε_g (Figure A-16a). Consider that the same material is now in a disordered state, such that all bonds remain satisfied, i.e., all atoms are surrounded by four nearest neighbors, so there are no dangling bonds. According to the Mott-CFO[*] model (see Refs. A5, A14, A19) the density-of-states curve is as shown in Figure A-16b. Note that the conduction and valence bands are broadened, and that the states in the tail regions are localized. This is in contrast to the states outside the tail, whose wave functions extend over some coherence distance §, which is long compared with the interatomic distance. The wave function of the localized state is centered on a lattice site, and falls sharply away from it (Figure A-17).

The disordered nature of the material introduces a randomness in the potential for the electrons. Theoretical models show that

---

[*]Mott, Cohen, Fritzsche, and Ovshinsky.

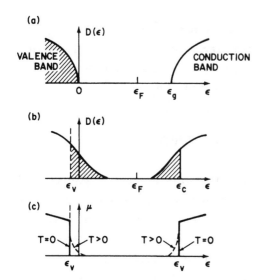

Figure A-16. (a) The density of states for a crystalline semiconductor with band gap, ϵ_g. (b) The same curve for a disordered (random) semiconductor. Localized states are shaded. (c) The variation of the mobility with energy. The mobility gap is $\epsilon_c - \epsilon_v$.

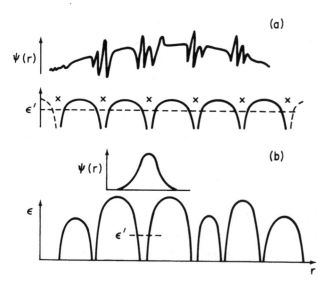

Figure A-17. (a) The periodic potential of a crystal and the extended wave function for an electron with energy ϵ'. (b) The potential of a disordered material and the localized wave function of an electron with energy ϵ'.

the energy region over which localized states exist increases as the strength of the randomness increases. Eventually, all states in the band become localized, which is the critical case discussed first by Anderson, and which is known as Anderson localization. (See Ref. A5.)

The mobility of the charge carriers is indicated in Figure A-16c. Any carrier between the energies ε_v and ε_c has zero mobility at T = 0K, giving rise to the term *mobility gap*. At finite temperature, some conduction can occur due to hopping, tunnelling, and other activation processes, to be discussed later.

The tails discussed above are not associated with specific defects, but with the randomness introduced into the potential by the random arrangement of atoms. Consider next the situation where specific defects are introduced in addition to this randomness, such as vacancies, dangling bonds, chain ends, etc. In these cases, extra states are introduced into the band gap, as illustrated in Figure A-18. There is discussion in the literature as to whether these extra states form one or two peaks. Two are shown in Figure A-18. In the case of defect states associated with the crystal (not impurities) these states are compensated, i.e., the lower peak will be filled, the higher empty. The presence of impurities can shift the Fermi energy, as drawn in Figure A-18 for acceptor impurities.

Simple models such as those discussed above cannot be applied to all cases. Models have not been theoretically developed for the different types of structural disorder that exist in carbons. By analogy with the models for Ge and Si, we may propose that the states

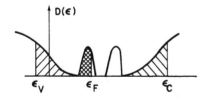

Figure A-18. The density-of-states curve for a disordered semiconductor with supplementary peaks for point defects, such as unsatisfied, or dangling, bonds. All states between ε_c and ε_v are localized. Shaded areas represent occupied states.

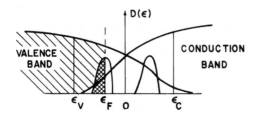

Figure A-19. A possible density-of-states curve for turbostratic graphite with other disorder, including dangling bonds, vacancies, etc. The energy zero is taken at the crossover between the valence and the conduction bands for defect-free 2D graphite. The Fermi energy is depressed so that hole conduction predominates. All states between ε_v and ε_c are localized.

near the band-overlap region in crystalline graphite become localized as a result of the disorder, as illustrated in Figure A-19. Also, subsidiary peaks in the density-of-states curves are associated with specific defects such as vacancies, dangling bonds, etc. The Fermi energy might be located at various positions throughout the region of localized states. It is possible that specific defects are not compensated in pregraphitic carbons. Thus, vacancies, for example, might act as acceptors, causing the Fermi energy to be lowered.

In the models developed for amorphous Ge and Si, conduction processes can be outlined as follows:

1. Band conduction involves electrons or holes excited out of the localized states into the extended states. The conductivity in this case will be controlled by the excitation of carriers into higher energy states and is of the form

$$\sigma = \sigma_0 \exp - \frac{\Delta\varepsilon_0}{kT} \qquad\qquad (A\text{-}39)$$

where $\Delta\varepsilon_0 = (\varepsilon_c - \varepsilon_f)$ for electrons, or $(\varepsilon_f - \varepsilon_v)$ for holes, depending on the predominant conduction type (i.e., position of ε_f); σ_0 is weakly dependent on T.

2. Thermally assisted tunnelling can occur in the mobility gap states. Normally, tunnelling to adjacent sites must be accompanied by a change in energy, since the probability

is low for a state with similar energy to be unoccupied in
the neighboring site. The conductivity is of the form

$$\sigma = \sigma_1 \exp - \frac{\Delta\epsilon_1}{kT} \qquad\qquad (A-40)$$

In this case $\sigma_1 \ll \sigma_0$, but $\Delta\epsilon_1 < \Delta\epsilon_0$ and is temperature
dependent. At lower temperature $\Delta\epsilon_1$ becomes smaller,
indicating that tunnelling is restricted increasingly to
states near ϵ_f. This leads to the third contribution.

3. Tunnelling near ϵ_f. The conductivity should have a form

$$\sigma = \sigma_2 \exp - \frac{\Delta\epsilon_2}{kT} \qquad\qquad (A-41)$$

where $\sigma_2 \ll \sigma_1$ and $\Delta\epsilon_2$ is the hopping energy, or change in
energy of the electron from the initial localized state to
the final. As the temperature is lowered it is increasingly
less probable for an electron to hop to the adjacent site,
since sufficiently energetic phonons are not available to
supply the necessary hopping energy. Instead hopping is
restricted to states within an energy ~kT of the initial
localized state, usually several atomic distances away.
A formula for this variable-rangehopping was developed by
Mott [A13].

$$\sigma = \text{constant} \quad \exp\left[-\left(\frac{T_0}{T}\right)^{1/4}\right] \qquad\qquad (A-42)$$

This holds for the three-dimensional case. For other dimensionality, n, where n = 1, 2, 3, the relevant formula is

$$\sigma = \text{constant} \quad \exp\left[-\left(\frac{T_0}{T}\right)^{1/n+1}\right] \qquad\qquad (A-43)$$

There are several types of carbon in which behavior of this type
has been seen.

Over an extended temperature range in disordered systems the conductivity is expected to change in character through the regimes
described. If typical Arrhenius plots are used, the activation energy

will not be constant, and it is often misleading to quote activation
energies in this case. At low temperature it is very difficult to
decide between behavior characterized by $\exp[-(T/T_0)^{1/4}]$ or
$\exp[-(T/T_0)^{1/3}]$, since the range of validity of Equation A-43 is
not clear.

It is again stressed that the models developed for three dimensional, disordered systems with tetrahedral bonds are being applied
to carbons by analogy, without theoretical justification. Much work
needs to be done in developing models for carbons and for relating
them to the variation of transport properties with position of the
Fermi level. At this point only a qualitative description of phenomena can be given.

ACKNOWLEDGMENTS

Thanks are due to many colleagues for discussing work included in
this review. Particular thanks are due to Professors S. Morozowski
and L. Vogel for reading the manuscript and for making many valuable
suggestions. Thanks are also due to Mrs. Florence Goldsworthy and
Mrs. Jeanne Fineran for typing the manuscript and Ms. Dawn Gillas
and Mr. Jan Scopel for assistance with numerous tasks in preparing
the manuscript. A grant from MERADCOM, Fort Belvoir for preparation
of figures is gratefully acknowledged, and also a Fellowship from
CENG, Grenoble, to the author during part of the time during which
the manuscript was in preparation.

REFERENCES

A. Books, Review Articles, and General Papers

A1. D. L. Beveridge and B. J. Bulkin, "Descriptive Crystal Orbital
    Theory of Conduction in Diamond and Graphite," *J. Chem. Educ.*,
    48, 587 (1971).

A2. L. C. F. Blackman (ed.), *Modern Aspects of Graphite Technology*,
    Academic, London and New York, 1970.

A3. L. Ebert, "Intercalation Compounds of Graphite," *Ann. Revs. Matls. Sci.*, 6, 181 (1976).

A4. A. R. Ford and E. Greenhalgh, "Industrial Applications of Carbon and Graphite," in *Modern Aspects of Graphite Technology* (L. C. F. Blackman, ed.), Academic, London and New York, 1970.

A5. H. Fritzsche, "Electronic Properties of Amorphous Semiconductors," in *Amorphous and Liquid Semiconductors* (J. Tauc, ed.), Plenum, London and New York, 1974.

A6. F. R. Gamble and T. H. Geballe, "Inclusion Compounds," in *Treatise on Solid State Chemistry*, Vol. III (N. B. Hannay, ed.), Plenum, New York, 1976, p. 89.

A7. E. P. Goodings, "Polymeric Conductors and Superconductors," *Endeavour* (U.K.), 34, 123 (1975).

A8. J. M. Hutcheon, "Polycrystalline Carbon and Graphite," in *Modern Aspects of Graphite Technology* (L. C. F. Blackman, ed.), Academic, London and New York, 1970.

A9. J. M. Hutcheon, "Manufacturing Technology of Baked and Graphitized Carbon Bodies," in *Modern Aspects of Graphite Technology* (L. C. F. Blackman, ed.), Academic, London and New York, 1970.

A10. D. W. McKee, "Carbon and Graphite Science," *Ann. Revs. Matls. Sci.*, 3, 195 (1973).

A11. K. Maschke and H. Overhof, "Influence of Stacking Disorder on the dc Conductivity of Layered Semiconductors," *Phys. Rev.*, B, 15, 2058 (1977).

A12. J. S. Miller and A. J. Epstein (eds.), "Synthesis and Properties of Low-Dimensional Materials," *Annals of New York Academy of Sciences*, Vol. 113, New York, 1978.

A13. N. F. Mott, "Electrons in Disordered Structures," *Advan. Phys.*, 16, 49 (1967).

A14. N. F. Mott and E. A. Davis, *Electronic Processes in Non-Crystalline Materials*, Clarendon, Oxford, 1971.

A15. A. Pacault, "Along the Carbon Way," *Carbon*, 12, 1 (1974).

A16. W. N. Reynolds, *Physical Properties of Graphite*, Elsevier, Amsterdam, 1968.

A17. J. A. Woollam, "Review of the Electrical and Thermal Properties of Graphite," *ACS Symposium Series* (1975) (M. L. Deviney and T. M. O'Grady, eds.), No. 21, p. 378.

A18. I. L. Spain, "The Electronic Properties of Graphite," in *Chemistry and Physics of Carbon*, Vol. 8, (P. L. Walker, Jr. and P. A. Thrower, eds.), Marcel Dekker, New York, 1973.

A19. J. Tauc (ed.), *Amorphous and Liquid Semiconductors*, Plenum, London and New York, 1974.

A20. Y. Toyozawa, "Theory of Localized Spins and Negative Magneto-
      resistance in Metallic Impurity Conduction," *J. Phys. Soc.
      Japan,* 17, 986 (1962).

A21. R. F. Trunin, et al., "Problem of the Existence of a Metallic
      State in Dynamically Compressed Carbon," *Sov. Phys. JETP,*
      29(4), 628 (1970).

A22. A. R. Ubbelohde and F. A. Lewis, *Graphite and Its Crystal Com-
      pounds,* Oxford, Oxford, 1960.

A23. D. W. Van Krevelen and J. Schuyer, *Coal Science,* Elsevier,
      1957.

A24. J. A. Van Vechten, "Quantum Dielectric Theory of Electronega-
      tivity in Covalent Systems III--Pressure-Temperature Phase
      Diagrams," *Phys. Rev.,* 7, B, 1479 (1973).

A25. J. M. Ziman, *Electrons and Phonons,* Clarendon, Oxford, 1962.

B. Selected Papers Relating to Structure of Carbons and Graphites

B1. S. Amelinckx, P. Delavignette, and M. Heerschap, "Dislocations
     and Stacking Faults in Graphite," Vol. 1, in *Chemistry and
     Physics of Carbon* (P. L. Walker, Jr., ed.), Marcel Dekker,
     New York, 1965.

B2. J. Biscoe and B. E. Warren, "An X-ray Study of Carbon Black,"
     J. Appl. Phys., 13, 364 (1942).

B3. J. D. Brooks and G. H. Taylor, "The Formation of Some Graphi-
     tizing Carbons," Vol. 4, in *Chemistry and Physics of Carbon*
     (P. L. Walker, Jr., ed.), Marcel Dekker, New York, 1968.

B4. D. B. Fischbach, "The Kinetics and Mechanism of Graphitization,"
     Vol. 7, in *Chemistry and Physics of Carbon* (P. L. Walker, Jr.,
     ed.), Marcel Dekker, New York, 1971.

B5. R. E. Franklin, "The Interpretation of Diffuse X-ray Diagrams
     of Carbon," *Acta Crystallogr.,* 3, 107 (1950).

B6. R. E. Franklin, "The Structure of Graphitic Compounds," *Acta
     Crystallogr.,* 4, 253 (1951).

B7. R. E. Franklin, "Crystallite Growth in Graphitizing and Non-
     graphitizing Carbons," *Proc. Roy. Soc.* (London), A, 209, 197
     (1951).

B8. A. A. Khomenko, et al., "On Interatomic Bonds in Transition
     Forms of Carbon," *Dokl. Akad. Nauk. SSR,* 206, 858-63 (1972).

B9. A. A. Khomenko, et al., "Thermal Transformation of Interatomic
     Bonds in Glassy Carbon," *Dokl. Akad. Nauk. SSR,* 206, 1112-15
     (1972).

B10. J. Maire and J. Mering, "Graphitization of Soft Carbons," Vol. 6, in *Chemistry and Physics of Carbon* (P. L. Walker, Jr., ed.), Marcel Dekker, New York, 1970.

B11. G. R. Millward and D. A. Jefferson, "Lattice Resolution of Carbons by Electron Microscopy," Vol. 14, in *Chemistry and Physics of Carbon* (P. L. Walker, Jr., ed.), Marcel Dekker, New York, 1978.

B12. I. Minkoff, "Graphite Crystallization," in *Preparation of Properties of Solid State Materials* (W. R. Wilcox, ed.), Marcel Dekker, New York, 1979.

B13. A. Pacault, "The Kinetics of Graphitization," Vol. 7, in *Chemistry and Physics of Carbon* (P. L. Walker, Jr., ed.), Marcel Dekker, New York, 1971.

B14. C. Roscoe and J. M. Thomas, "The Identification and Some Physio-chemical Consequences of Non-basal Edge and Screw Dislocations in Graphite," *Proc. Roy. Soc.*, 297, 397 (1966).

B15. W. Ruland, "X-ray Diffraction Studies on Carbon and Graphite," Vol. 4, in *Chemistry and Physics of Carbon* (P. L. Walker, Jr., ed.), Marcel Dekker, New York, 1968.

B16. P. A. Thrower, "The Study of Defects in Graphite by Transmission Electron Microscopy," Vol. 5, in *Chemistry and Physics of Carbon* (P. L. Walker, Jr., ed.), Marcel Dekker, New York, 1969.

C. Selected Papers in Energy Bands and Bonding in Graphite, Diamond, and Carbons[*]

C1. U. Berg, G. Drager, and O. Brummer, "Combined Investigations of the Valence Band Structure of Graphite by K X-ray Emission Spectroscopy and X-ray Photoemission Spectroscopy," *Phys. Status Solidi* (Germany), B74, 341 (1976).

C2. A. D. Boardman, M. I. Darby, and E. T. Micah, "Density of States in Perturbed Graphite Lattice," *Carbon*, 11, 207 (1973).

C3. C. A. Coulson, "Electronic Structure of the Boundary Atoms of a Graphite Layer," *Proc. Con. Carbon*, 215 (1960).

C4. C. A. Coulson, L. J. Schaad, and L. Burnelle, "Benzene to Graphite--The Change in Electronic Energy Levels," *Proc. Con. Carbon*, 27 (1957).

C5. R. O. Dillon, I. L. Spain, and J. W. McClure, "Electronic Energy Band Parameters of Graphite and Their Dependence on Pressure, Temperature, and Acceptor Concentration," *J. Phys. Chem. Solids*, 38, 635 (1977).

---

[*]See also A1, A18, D12, E16.

C6.   E. Doni and G. Pastori Parravicini, "Energy Bands and Optical Properties of Hexagonal BN and Graphite," *Nuovo Cimento*, B, LXIV, 117 (1969).

C7.   Y. Fukuda, "Electronic States in Perturbed Two-dimensional Graphite Lattice," *J. Phys. Soc. Japan*, 20, 353 (1965).

C8.   R. R. Haering and P. R. Wallace, "Electric and Magnetic Properties of Graphite," *J. Phys. Chem. Solids*, 3, 253 (1957).

C9.   R. R. Haering and S. Mrozowski, "Band Structure and Electronic Properties of Graphite Crystals," *Prog. Semiconductors*, V, 273 (1960).

C10. C. Hamann, "Electronic Properties of Molecular Crystals," *Krist. Tech.* (Germany), 12, 651 (1977).

C11. L. A. Hemstreet, C. Y. Fong, and M. L. Cohen, "Calculation of the Band Structure and Optical Constants of Diamond Using the Nonlocal-Pseudopotential Method," *Phys. Rev.*, B, 2, 2054 (1970).

C12. G. A. Kapustin and E. Z. Meilikhov, "Metal-Semiconductor and Semiconductor-to-metal Transitions in Graphite in a Magnetic Field," *Fiz Tverd Tela* (USSR), 17, 2978 (1975) [trans. *Sov. Phys. Solid State*, 17, 1979 (1975)].

C13. J. R. Leite, B. I. Bennett, and F. Herman, "Electronic Structure of the Diamond Crystal Based on an Improved Cellular Calculation," *Phys. Rev.*, B, 12, 1466 (1975).

C14. A. Marchand, "Sur une Modification du modele de Slonczewski et Weiss Applicable aux Curbones Pregraphitiques," *Compt. Rend.*, 256, 3070 (1963).

C15. J. W. McClure, "Electron Energy Band Structure and Electronic Properties of Rhombohedral Graphite," *Carbon*, 7, 425 (1969).

C16. J. W. McClure, and J. Ruvalds, "Energy Band Structure and Electronic Properties of Turbostratic Graphite," submitted to *Phys. Rev.*

C17. J. W. McClure, "Energy Band Structure of Graphite," *Phys. of Semimetals and Narrow Gap Semiconductors* (Carter and Bate, eds.) (1971), p. 127.

C18. J. W. McClure, "Electronic Structure and Magnetic Properties of Monocrystalline Graphite," *J. Chim. Phys*, 859 (1961).

C19. I. A. Misurkin and A. A. Ouchinnikou, "Electronic Structure of High Π-electron Systems (Graphite, Polyacene, Cumulene)," *Theor. Chim. Acta*, 13, 115 (1969).

C20. S. Mrozowski, "Zone Structure of Graphite," *Phys. Rev.*, 92, 1320 (1953).

C21. H. Nagashi, M. Tsukada, K. Nakao, and Y. Uemura, "Combined OPW-TB Method for the Band Calculation of Layer-Type Crystals II--The Band Structure of Graphite," *J. Phys. Soc. Japan*, 35, 396 (1973).

C22. H. Nagayoshi, K. Nakao, and Y. Uemura, "Band Theory of Graphite I--Formalism of a New Method of Calculation and the Fermi Surface of Graphite," *J. Phys. Soc. Japan*, 41, 1480 (1976).

C23. H. Nagayoshi, K. Nakao, and Y. Uemura, "Fermi Surface Parameters of Graphite by Band Calculations," *Sol. St. Comm.* (U.S.), 18, 225 (1976).

C24. H. Nagayoshi, "Band Theory of Graphite II--Pressure Effects on the Fermi Surface," *J. Phys. Soc. Japan*, 43, 760 (1977).

C25. K. Nakao, "Landau Level Structure and Magnetic Breakdown in Graphite," *J. Phys. Soc. Japan*, 40, 761 (1976).

C26. M. Nishida, "Application of Molecular Orbital Method to Crystalline Solids: Calculation of the Electronic Energy Bands of Diamond-Type Crystals," *J. Chem. Phys.*, 69, 956 (1978).

C27. T. Shimizu and N. Ishii, "Energy Bands and Their Pressure Dependence of Diamond- and Zincblende-Type Crystals by Molecular Orbital Method," *Mem. Fac. Technol. Kanazawa Univ.* (Japan), 11, 11 (1977).

C28. J. C. Slonczewski and P. R. Weiss, "Band Structure of Graphite," *Phys. Rev.*, 109, 272 (1958).

C29. J. M. Thomas, E. L. Evans, M. Barber, and P. Swift, "Determination of the Occupancy of Valence Bands in Graphite, Diamond and Less-ordered Carbons by X-ray Photo-electron Spectroscopy," *Trans. Faraday Soc.*, 67(583), 1875 (1971).

C30. W. W. Toy, C. R. Hewes, and M. S. Dresselhaus, "Magnetoreflection studies of Single Crystal Pyrolytic and Kish Graphite," *Carbon*, 11, 575 (1973).

C31. M. Tsukada, K. Nakao, Y. Uemura, and S. Nagai, "Combined OPW-TB Method for the Band Calculation of Layer Type Crystals I--General Formalism and Application to the π-Band of Graphite," *J. Phys. Soc. Japan*, 32, 54 (1972).

C32. P. R. Wallace, "The Band Theory of Graphite," *Phys. Rev.*, 71, 622 (1947).

C33. W. Wegener and L. Fritsche, "A Study of the Band Structure of Graphite Based on the Rigorous Cellular Method," *Phys. Status Solidi*, B (Germany), 78, 585 (1976).

C34. A. Zunger, "Self-Consistent LCAO Calculation of the Electronic Properties of Graphite. I. The Regular Graphite Lattice," *Phys. Rev.*, B, 17, 626 (1978).

C35. A. Zunger and R. Englman, "Self-Consistent LCAO Calculation of the Electronic Properties of Graphite. II. Point Vacancy in the Two-Dimensional Crystal," *Phys. Rev.*, B, 17, 642 (1978).

C36. J. Zupan, "Energy Bands in Boron Nitride and Graphite," *Phys. Rev.*, B, 6, 2477 (1972).

D. General and Theoretical Papers on Electrical
   Transport of Carbons and Graphite

D1. A. D. Boardman and D. G. Graham, "Residual Resistivity in
     Ideal Graphite," *J. Phys. Chem. (Sol. St. Phys.)*, Ser. 2, 2,
     2320 (1969).

D2. P. Delhaes, "Positive and Negative Magnetoresistance in Carbons,"
     in *Chemistry and Physics of Carbon*, Vol. 7 (P. L. Walker, Jr.,
     ed.), Marcel Dekker, New York, 1971.

D3. R. O. Dillon and I. L. Spain, "Galvanomagnetic Effects in
     Graphite. II--The Influence of Trigonal Warping of the Con-
     stant Energy Surfaces on $\sigma_{xx}(B)$," *J. Phys. Chem. Solids* (U.K.),
     39, 923 (1978).

D4. R. O. Dillon and I. L. Spain, "Effects of Warped Energy Sur-
     faces on the Low Field Hall Coefficient of Graphite, *Solid
     State Comm.* (U.K.), 26, 333 (1978).

D5. T. Hirai, S. Yajima, and K. Murakami, "A Vector--Vector Effect
     in the Electrical Conductivity of Graphite," *Carbon*, 7, 625
     (1969).

D6. Y. Hishiyama, "Magnetoresistivity as a parameter for Determina-
     tion of the Degree of Graphitization," *Carbon*, 13, 244 (1975).

D7. V. V. Kechin, "On the Theory of Galvanomagnetic Effects in
     Graphite," *Fiz. Tverd. Tela*, 11(7), 1788 (1969).

D8. V. V. Kechin, "Theory of the Galvanomagnetic Effects in
     Graphite," *Sov. Phys. Solid State*, 11(7), 1448 (1970).

D9. T. Kamura and K. Yazawa, "Explanation of the Anomalous Magneto-
     resistance of Carbon," *Denki Tsushin Dougakn Gakuho*, 25(2),
     223 (1975).

D10. J. W. McClure, "Field Dependence of Magnetoconductivity,"
     Phys. Rev., 101, 1642 (1956).

D11. J. W. McClure, "Analysis of Multicarrier Galvanomagnetic Data
     for Graphite," *Phys. Rev.*, 112, 715 (1958).

D12. J. W. McClure, "Relation between Electron Energy Band Structure
     and the Properties of Graphite," *Proc. Conf. Carbon 1960*,
     Pergamon, New York.

D13. J. W. McClure and L. B. Smith, "Theory of the Electron Trans-
     port of Single Crystalline Graphite," *Proc. 5th Carbon Conf.*,
     2, 3 (1961).

D14. K. Noto and T. Tsuzuku, "A Simple Two-band Model of Galvano-
     magnetic Effects in Graphite in Relation to the Magnetic Field
     Azimuth," *Jpn. J. Appl. Phys.*, 14, 46 (1975).

D15. K. Noto and T. Tsuzuku, "The Transverse Magnetoresistance of
     Graphite in Relation to the Magnetic Field Azimuth," *J. Phys.
     Soc. Japan*, 35, 1264 (1973).

D16. K. Noto and T. Tsuzuku, "Kohler's Rule for Magnetoresistance of Graphite and Carbons," *Carbon*, 12, 209 (1974).

D17. K. Noto and T. Tsuzuku, "Magnetoresistance of Graphite: Deviation from the H^2 Law and Evaluation of Conduction Parameters," *Tanso*, 84, 20 (1976).

D18. S. Ono and K. Sugihara, "Theory of the Transport Properties in Graphite," *J. Phys. Soc. Japan*, 21, 861 (1966).

D19. S. Ono and K. Sugihara, "Trigonal Warping of the Bands and Hall Effect in Graphite," *J. Phys. Soc. Japan*, 24, 818 (1968).

D20. L. A. Pesin, P. V. Pekin, and V. Yu. Karasco, "Temperature Dependence of the Specific Electric Resistivity of Carbon Materials," *Vopr. Fiz. Tver. Tela*, 7, 39 (1977).

D21. V. I. Petravichev, V. P. Sobalev, and N. P. Kiselev, "Correlation between the Coefficient of Thermal Conductivity and the Coefficient of Electrical Conductivity of Graphite," *Vopr. Teplofiz. Yader. Reaktorov*, 5, 71 (1976).

D22. U. Pospelov and V. V. Kechin, "Dependence of the Electrical Resistivity of Graphite on the Magnetic Field," *Fiz. Tverd. Tela*, 5, 3574 (1963).

D23. Y. A. Pospelov, "On the Theory of Graphite Galvanomagnetic Properties," *Fiz. Tverd. Tela*, 12(3), 835 (1970).

D24. Y. A. Pospelov, "Theory of the Electrical Resistivity of Graphite," *Fiz. Tverd. Tela* (USSR), 13, 2314 (1971) [trans. *Sov. Phys. Solid State* (U.S.), 13 (1971)].

D25. I. L. Spain, "Galvanomagnetic Data and the Sign of the Carriers Located at Point K in the Brillouin Zone of Graphite," *J. Chem. Phys.*, 52, 5, 2763 (1970).

D26. I. L. Spain and R. O. Dillon, "Kohler's Rule and Other Scaling Relationships for the Magnetoresistance of Graphite," *Carbon*, 14, 23 (1976).

D27. K. Sugihara and H. Sato, "Electrical Conductivity of Grahpite," *J. Phys. Soc. Japan*, 18(3), 332 (1963).

D28. K. Sugihara, "Effect of Carrier-Carrier Scattering on the Hall Effect in Graphite," *J. Phys. Soc. Japan*, Suppl. 21, 324 (1966).

D29. K. Sugihara and S. Ono, "Galvanomagnetic Properties of Graphite at Low Temperatures," *J. Phys. Soc. Japan*, 21, 631 (1966).

D30. J. A. Woollam, "Spin Splitting, Fermi Energy Changes and Anomalous g Shifts in Single-crystal and Pyrolytic Graphite," *Phys. Rev. Lett.*, 25(12), 810 (1970).

D31. J. A. Woollam, "Direct Evidence for Majority Carrier Locations in the Brillouin Zone of Graphite," *Phys. Lett.*, 32(2), 115 (1970).

D32. J. A. Woollam, "Graphite Minority Carriers," *Phys. Lett.*, A, 35(5), 332 (1971).

D33. J. A. Woollam, "Minority Carriers in Graphite," *Phys. Rev.*, B, 4(10), 3393 (1971).

D34. H. Yasunaga, N. Okyuama, and K. Takaya, "Plane Hall Effect in Pyrolytic Graphite," *J. Phys. Soc. Japan*, 26(4), 1062 (1969).

D35. D. A. Young, "The Role of Localized Electronic States in Disordered Carbon and Graphite," *Carbon*, 6, 135 (1968).

E. Natural Single Crystal and Kish Graphite[*]

E1. G. Alquie and A. Kreisler, "Low-Temperature Microwave Absorption in Natural and Pyrolytic Graphites," *Phys. Status Solidi*, A (Germany), 29, 77 (1975).

E2. S. B. Austerman, "Growth of Graphite Crystals from Solution," in *Chemistry and Physics of Carbon*, Vol. 4 (P. L. Walker, Jr., ed.), Marcel Dekker, New York, 1968.

E3. S. B. Austerman, S. M. Myross, and J. W. Wagner, "Growth and Characterization of Graphite Single Crystals," *Carbon*, 5, 549 (1967).

E4. T. G. Berlingcourt and M. C. Steele, "Oscillatory Hall Effect, Magnetoresistance and Magnetic Susceptibility of a Graphite Single Crystal," *Phys. Rev.*, 98(4), 956 (1955).

E5. R. Bhattacharya, "Effect of Chemical Treatment on the Electrical Conductivity of Graphite," *Indian J. Phys.*, 33, 407 (1959).

E6. R. Bhattacharya, "Magnetoresistance in Single Crystals of Graphite," *Indian J. Phys.*, 34, 53 (1965).

E7. R. Bhattacharya and A. K. Dutta, "Misalignments in Natural Graphite Crystals and Their Possible Effects on its Magnetic and Electrical Properties," *Indian J. Phys.*, 50, 495 (1976).

E8. A. K. Dutta, "Electrical Conductivity of Single Crystals of Graphite," *Phys. Rev.*, 90, 187 (1953).

E9. K. Kawamura, N. Sato, T. Aoki, and T. Tsuzuku, "Hall Coefficient of Graphite in the Magnetic Field Range Below 1.2 kOe," *J. Phys. Soc. Japan*, 41(6), 2027 (1976).

E10. K. Kawamura, T. Saito, and T. Tsuzuku, "Temperature Dependence of the Galvanomagnetic Properties of Graphite Between 4.2 K and 298 K," *J. Phys. Soc. Japan*, 42(2), 574 (1977).

E11. G. H. Kinchin, "The Electrical Properties of Graphite," *Proc. Roy. Soc.*, A, 217, 9 (1953).

---

[*]See also A18, M13, M14, N12, O3, O12, O14, T4, T7, T8.

E12. W. Primak and L. H. Fuchs, "Electrical Conductivities of
      Natural Graphite Crystals," *Phys. Rev.*, 95, 22 (1954).

E13. S. Ray and R. Bhattacharya, "Quantitative Estimation of Mis-
      alignment of the Layers in Natural Crystals of Graphite,"
      Indian J. Phys., 33, 407 (1959).

E14. D. E. Soule, "Analysis of Galvanomagnetic de Hass-van Alphen
      Type Oscillations in Graphite," *Phys. Rev.*, 112, 708 (1958).

E15. D. E. Soule, "Magnetic Field Dependence of the Hall Effect and
      Magnetoresistance in Graphite Single Crystals," *Phys. Rev.*,
      112, 698 (1958).

E16. D. E. Soule and J. W. McClure, "Band Structure and Transport
      Properties of Single-Crystal Graphite," *J. Phys. Chem. Sol.*,
      8, 29 (1959).

E17. D. E. Soule, J. W. McClure, and L. B. Smith, "Study of the
      Shubnikov-de Haas Effect. Determination of the Fermi Surfaces
      in Graphite," *Phys. Rev.*, 134, A, 453 (1964).

E18. G. E. Washburn, "Der Einfluss der Magnetisierung auf den
      Gleichstromwiderstand des Graphits nach der Hauptachse,"
      Ann. Phys. (Leipzig), 48, 236 (1915).

F. Highly Oriented Pyrolytic Graphite[*]

F1. J. D. Cooper, J. Woore, and D. A. Young, "Electronic Properties
     of Well Oriented Graphite," *Nature*, 225, 5234, 721-722 (1970).

F2. R. O. Dillon, I. L. Spain, J. A. Woollam, and W. H. Lowrey,
     "Galvanomagnetic Effects in Graphite I: Low Field Data and
     the Densities of Free Carriers," *J. Phys. Chem. Solids*, 39,
     907 (1978).

F3. J. F. Green, P. Bolsaitis, and I. L. Spain, "Pressure Dependence
     of C-Axis Elastic Parameters of Oriented Graphite," *J. Phys.
     Chem. Solids*, 34, 1927 (1973).

F4. Y. Hishiyama, A. Ono, T. Tsuzuku, and T. Takeyawa, "Galvano-
     magnetic Properties of Well-oriented Graphite in Relation to
     the Structural Imperfections," *Jpn. J. Appl. Phys.*, 11(7), 958
     (1972).

F5. A. W. Moore, A. R. Ubbelohde, and D. A. Young, "Stress Recrystal-
     lization of Pyrolytic Graphite," in *Chemistry and Physics of
     Carbon*, Vol. 11 (P. L. Walker, Jr. and P. A. Thrower, eds.),
     Marcel Dekker, New York, 1973, p. 69.

[*]See also A18, N5, N6, O3, O13, O15, O16, T3, T7, T8.

F6. A. W. Moore, A. R. Ubbelohde, and D. A. Young, "Stress Recrys-
     talization of Pyrolytic Graphite," *Proc. Roy. Soc.*, A, 280,
     153 (1964).

F7. I. L. Spain, A. R. Ubbelohde, and D. A. Young, "Anisotropy of
     Electrical Properties of Well-Oriented Graphites," *Soc. Chem.
     Ind.*, 123 (1966).

F8. I. L. Spain, A. R. Ubbelohde, and D. A. Young, "Electronic
     Properties of Well-Oriented Graphite," *Phil. Trans. Roy. Soc.*
     (London) 262, 345 (1967).

F9. C. Zeller, G. M. T. Foley, and F. L. Vogel, "The Effect of
     Extrinsic Defects in Pyrolytic Graphite on the A-Axis Resis-
     tivity," *J. Mater. Sci.*, 13, 1114 (1978).

G. Pyrolytic Carbons and Graphites[*]

G1. J. C. Bokros, "Deposition Structure and Properties of Pyrolytic
     Carbon," in *Chemistry and Physics of Carbon*, Vol. 5 (P. L.
     Walker, Jr., ed.), Marcel Dekker, New York, 1969.

G2. A. R. G. Brown, A. R. Hall, and W. Watt, "Density of Deposited
     Carbon," *Nature*, 172, 1145 (1953).

G3. L. C. Blackman, G. Saunders, and A. R. Ubbelohde, "Defect
     Structure and Properties of Pyrolytic Carbons," *Proc. Roy.
     Soc.*, A, 264, 19 (1961).

G4. V. Ya Chekhovski, V. A. Petrov, and I. L. Petrova, "Effect of
     Heat Treatment Temperature on Thermal Conductivity and Elec-
     trical Resistivity of Pyrographite," *Teplofiz. Vys. Temp.*
     (USSR), 9, 851 (1971) [trans. *High Temp.* (U.S.), 9, 771 (1971)].

G5. P. de Kepper, P. Delhaes, and H. Gasparoux, "Comparative Evolu-
     tion of the Magnetic Anisotropy and Magnetoresistance of a
     Family of Pyrocarbons During Graphitization," *Comp. Rend. Hebd.
     Seances Acad. Sci.* (France), C, 276, 1369 (1973).

G6. P. Delhaes, H. Gasparoux, and M. Uhlrich, "Anisotropic Negative
     Magnetoresistance in a Pyrolytic Carbon," *Phys. Lett.*, A, 34,
     417 (1971).

G7. P. Delhaes, P. de Kepper, and M. Uhlrich, "A Study of the Nega-
     tive Magnetoresistance of Pyrocarbons," *Phil. Mag.*, 29, 1301
     (1974).

G8. A. V. Dmitriev and S. V. Shulepco, "Temperature Dependence of
     the Specific Electric Resistance of Carbon-Graphite Materials,"
     Vopr. Fiz. Tverd. Tela, 7, 63 (1977).

---

[*]See also E1, M3, M8, M15, M17-M20, N2, N9, O9, X22.

G9.   D. B. Fischbach, "Preferred Orientation Parameters for Pyro-
      lytic Carbons," *J. Appl. Phys.*, 37, 2202 (1966).

G10. S. Gromb, "Etude de Proprietes Electroniques de Pyrocarbones
      et de leur Variation Thermique," *Comp. Rend.*, 256, 4002 (1963).

G11. O. J. Guentert and C. A. Klein, "Preferred Orientation and
      Anisotropy Ratio of Pyrolytic Graphite," *Appl. Phys. Lett.*,
      2, 125 (1963).

G12. C. A. Klein, "Electrical Properties of Pyrolytic Graphites,"
      Rev. Mod. Phys., 34, 56 (1962).

G13. C. A. Klein, "Pyrolytic Graphites, Their Description as Semi-
      metallic Molecular Solics," *J. Appl. Phys.*, 33, 3338 (1962).

G14. C. A. Klein, "STB Model and Transport Properties of Pyrolytic
      Graphites," *J. Appl. Phys.*, 35, 2947 (1964).

G15. C. A. Klein, "Electronic Transport in Pyrolytic Graphite and
      Boron Alloys of Pyrolytic Graphite," in *Chemistry and Physics
      of Carbon*, Vol. 2 (P. L. Walker, Jr., ed.), Marcel Dekker,
      New York, 1966.

G16. A. S. Kotosonov, V. A. Vinnikov, A. I. Polozhikin, V. I.
      Frolov, and V. P. Sosedov, "Effect of Chlorine on the Changes
      of Electronic Properties of Pyrolytic Carbon in the Course of
      Graphitization," *Carbon*, 8, 389 (1970).

G17. A. I. Lutkov, V. I. Volga, B. K. Dymov, E. Yu Lukina, and
      P. V. Tamarin, "Thermal and Electric Properties of Pyrolytic
      Graphite," *Izv. Akad. Nauk SSSR, Neorgan. Mater.*, 8, 1409
      (1972) [trans. *Inorg. Mater.* (U.S.), 8, 1210 (1972)].

G18. A. I. Lutkov, V. I. Volga, and B. K. Dymov, "Methods of
      Determining the Average Size of Graphite Crystallites in the
      Basal Plane," *Zavod. Lab.* (USSR), 39, 1201 (1974) [trans.
      Indust. Lab. (U.S.), 39, 1570 (1974)].

G19. A. W. Moore, "The Induction Heating of Pyrolytic Graphite,"
      Carbon, 5, 159 (1967).

G20. R. A. Morant, "Some Properties of Resistance-grown Pyrolytic
      Graphite," *Brit. J. Appl. Phys.*, 17, 75 (1966).

G21. R. A. Morant, "Crystallite Size of Pyrolytic Graphite," *J.
      Phys. D. Appl. Phys.*, 3, 1367 (1970).

G22. V. A. Petrov, I. I. Petrova, V. Ya. Chekovskoi, and E. N.
      Lyukshin, "Specific Electrical Resistivity of Pyrographite,"
      Teplofiz. Vys. Temp. (USSR), 9, 302 (1971) [trans. *High Temp.*
      (U.S.), 9, 271 (1971)].

G23. A. I. Polozhikhin and A. S. Kotoscnov, "Effect of Fast Cooling
      on the Galvanomagnetic Effects in Synthetic Graphite," *Khim.
      Tverd. Topl.* (Moscow), 1, 123 (1976).

G24. A. R. Saha, P. K. Banerjee, and A. K. Das, "On the Electrical
      Resistivities of Pyrolytic Graphite," *Indian J. Phys.*, 44,
      438 (1970).

G25.   A. R. Saha, P. K. Banerjee, and A. K. Das, "Propagation of
       Electromagnetic Waves in Pyrolytic Graphite," *Proc. IEEE*,
       60(1), 140 (1972).

G26.   G. Saunders, "Negative Magnetoresistance in Pyrolytic Carbons
       and Graphite Bromine," *Appl. Phys. Lett.*, 4, 138 (1964).

G27.   K. Takeya and K. Yazawa, "Unusual Galvanomagnetic Properties
       of Pyrolytic Graphite," *J. Phys. Soc. Japan*, 19, 138 (1964).

G28.   K. Takeya, K. Yazawa, F. Ezoe, N. Okuyama, and H. Akutsu,
       "Unique Behaviors in the Oscillatory Magneto-Resistance of
       Pyrolytic Carbons at 77K," *J. Phys. Soc. Japan*, 20, 1735 (1965).

G29.   K. Takeya, K. Yazawa, N. Okuyama, and H. Akutsu, "Evidence
       for the Existence of Extremely Light Carriers in Pyrolytic
       Carbons," *Phys. Rev. Lett.*, 15, 111 (1965).

G30.   K. Takeya, K. Yazawa, N. Okuyama, H. Akutsu, and F. Ezoe,
       "Shubnikov-de Haas Phenomena in Pyrolytic Carbons at Liquid
       Nitrogen Temperature," *Phys. Rev. Lett.*, 15, 110 (1965).

G31.   K. Yazawa, "Hall Effect of Pyrolytic Carbons," *J. Chim. Phys.*
       (France), 64, 961 (1967).

G32.   K. Yazawa, "Negative Magneto-Resistance in Pyrolytic Carbons,"
       J. Phys. Soc. Japan, 26, 1407-1419 (1969).

G33.   Yu. I. Sementsov, E. I. Khar'kov, and L. Yu. Vavilina, "Effect
       of Grain Boundary Scattering on Temperature Dependence of
       Electroresistivity of Nonequilibrium Graphites," *Ukr. Fiz. Zh.*
       (USSR), 23, 616 (1978).

H.   Graphitizable Carbons[*]

H1.   F. Boy and A. Marchand, "Validite de Quelques Modeles Electron-
       iques Simple de Carbones," *Carbon*, 5, 227 (1967).

H2.   F. Carmona, P. Delhaes, G. Keryer, and J. P. Manceau, "Non-
       Metal to Metal Transition in a Non-Crystalline Carbon," *Solid
       State Comm.*, 14, 1183 (1974).

H3.   V. V. Galipern and Y. M. Obukhowskii, "Specific Electrical
       Resistance of Coke, Semi-Coke, Thermo-Anthracite, and Graphite
       Powders Studied by a Two-Probe Method," *Khim. Tverd. Topl.*, 2,
       121-128 (1973).

H4.   O. K. Griffith, Jr. and R. I. Gayley, "The Temperature Depend-
       ence of the Electrical Resistivity of Soft Carbon Below 4.2 K--
       Letter to the Editor," *Carbon*, 3, 541 (1966).

---

[*]See also M1, M2, M4, M7, M10-M12, N8, N20, N21.

H5. O. Hauser, "On the Graphitization of Carbons--Hall Effect and
     Resistivity," *J. Phys. Chem.*, 210, 151 (1958).

H6. Y. Hishiyama, "Negative Magnetoresistance in Soft Carbons and
     Graphite," *Carbon*, 8, 259 (1970).

H7. Y. Hishiyama, et al., "Field Dependence of Magnetoresistance
     in no-binder Soft Carbons at 77 K," *Jpn. J. Appl. Phys.*, 10,
     416 and 820 (1971).

H8. H. Honda, K. Egi, S. Toyoda, Y. Sanda, and T. Furuta, "Elec-
     tronic Properties of Heat-treated Coals," *Carbon*, 1, 155 (1964).

H9. M. Inagaki, Y. Komatsu, J. V. Zanchetta, "Hall Coefficient and
     Magnetoresistance of Carbons and Polycrystalline Graphite in
     the Temperature Range 1.5-300 K," *Carbon*, 7, 163 (1969).

H10. T. Kimura and K. Yazawa, "Evaluation of Energy Band Parameters
     of Carbons," *Carbon*, 11, 139 (1973).

H11. E. E. Loebner, "Thermoelectric Power, Electrical Resistance,
     and Crystalline Structure of Carbons," *Phys. Rev.*, 102, 46
     (1956).

H12. J. Millet, J. Rogue, A. Vivones, A. Descampo, and J. Millet,
     "Modification des Proprietes Semiconductrices des Carbones au
     Cours de la Graphitization," *J. Chim. Phys.*, 62(1), 46 (1965).

H13. S. Mrozowski, "Electrical Resistivity of Polycrystalline
     Graphite and Carbons," *Phys. Rev.*, 77, 838 (1950).

H14. S. Mrozowski, "Semiconductivity and Diamagnetism of Poly-
     crystalline Graphite and Condensed Ring Systems," *Phys. Rev.*,
     85, 609 (1952).

H15. S. Mrozowski, "The Nature of Artificial Carbons," *Soc. Chem.
     Ind.* (U.K.), 7 (1958).

H16. S. Mrozowski, "Electron Spin Resonance in Turbocrystalline
     Carbons--I," *Carbon*, 6, 243 (1968).

H17. S. Mrozowski, "Electronic Properties and Band Model of Carbons,"
     Carbon, 9, 97 (1971).

H18. S. Mrozowski and A. Chaberski, "Hall Effect and Magnetoresis-
     tivity in Carbons and Polycrystalline Graphites," *Phys. Rev.*,
     104, 74 (1956).

H19. M. Nakamizo, H. Honda, and M. Inagaki, "Raman Spectra, Effec-
     tive Debye Parameter and Magnetoresistance of Graphitized
     Cokes," *Carbon*, 15, 295 (1977).

H20. E. I. Parnov and R. B. Orshanskii, "Measurement of the Elec-
     trical Resistivity of Carbon Materials at High Temperatures,"
     Zavod. Lab., 29, 1112 (1963).

H21. Z. I. Syumycov, M. F. Goliakbarov, and R. A. Gizatullin,
     "Variation of the Specific Electrical Conductivity of Petro-
     leum Cokes During Roasting," *Khim. Tekhnol. Topl. Masel*, 4,
     35 (1965).

H22. I. M. Yarmolanka, "Electrical Conductivity of Activated Carbon Materials," *Vestn. Akad. Nauk BSSR, Ser. Khim. Nauk*, 1, 52 (1972).

H23. F. I. Zorin and A. S. Kotosonov, "Hall Effect and Electrical Conductivity of Pyrocarbon Specimens Deposited at 2000°C and 2200°C," *Izv. Akad. Nauk SSSR, Ser. Neorg. Mater.*, 13, 539 (1977).

I. Bonded Polycrystalline Carbons and Graphites[*]

I1. S. G. Bapat, "Thermal Conductivity and Electrical Resistivity of Two Types of ATJ-S Graphite to 3500 K," *Carbon*, 11, 511 (1973).

I2. S. G. Bapat and H. Nickel, "Thermal Conductivity and Electrical Resistivity of Poco Grade AXF-Q1 Graphite to 3300 K," *Carbon*, 11, 323 (1973).

I3. C. Bhatia, R. K. Aggarwal, and P. Ranjan, "A Simple Method for the Estimation of Optimum Binder Content in Baked Carbon Mixes," *J. Mater. Sci.* (U.K.), 12, 1639 (1977).

I4. B. K. Dymov, A. I. Lutkov, and V. I. Volga, "Thermal and Electrical Conductivity of Graphite Obtained by Thermomechanical Processing in the 80-2500 K Range," *Heat Transfer--Soviet Res.* (U.S.), 5, 149 (1973).

I5. M. Eto and T. Oku, "Change in Electrical Resistivity of Nuclear Graphite During Compression Tests and a Model for its Deformation and Fracture Mechanism," *J. Nucl. Mater.*, 54, 245 (1974).

I6. M. Eto, T. Usin, and T. Oku, "Change in Electrical Resistance of Nuclear Graphite during Compressive Tests--Letter to the Editor," *J. Nucl. Mater.*, 45, 347 (1972).

I7. V. S. Glazachev, V. S. Bokunov, D. N. Polubay, and S. M. Rabinovich, "Temperature Dependence of Electric Conductivity of Some Carbon-Graphite Materials," *Izv. Akad. Nauk SSSR, Ser. Neorg. Mater.*, 12(3), 544 (1976).

I8. K. J. Huttinger, "Discontinuities of the Electrical Resistivity of Young's Modulus of Pitch Bonded Green Carbon Artifacts in the HTT Range 400-500°C--Letters to the Editor," *Carbon*, 9, 809 (1971).

I9. A. I. Lutcov, V. I. Volga, and B. K. Dymov, "Thermal Conductivity, Electric Resistivity and Specific Heat of Dense Graphites," *Carbon*, 8, 753 (1970).

---

[*]See also A8, A9, E11, H3, H9, H13, H18, M5, M16, N1, N13, N14, N17, N19, N23.

I10. A. I. Lutcov, B. K. Dymov, V. I. Volga, "Relation Between
      Thermal and Electrical Resistivities of Graphite," *Inzh.-Fiz.
      Zh.* (USSR), 22, 932 (1972) [trans. *J. Eng. Phys.* (U.S.), 22,
      654 (1972)].

I11. H. Matsuo, "Relationship between Electrical Resistivity and
      Thermal Conductivity of Reactor Grade Graphite Irradiated with
      Neutrons at 300°C-400°C—Letter to the Editor," *J. Nucl. Mater.*,
      42, 105 (1972).

I12. H. Matsuo and T. Honda, "Annealing of Reactor-Grade Graphite
      Irradiated with Neutrons at 350°C; Effect on Electrical Resis-
      tance and Thermal Conductivity--Letter to the Editor," *J. Nucl.
      Mater.*, 45, 79 (1972).

I13. M. L. Minges, "Analysis of Thermal and Electrical Energy Trans-
      port in Poco AXM-5Q1 Graphite," *Int. J. Heat Mass Transfer*
      (U.K.), 20, 1161 (1977).

I14. G. L. Montet, "Low Temperature Galvanomagentic Properties of
      Graphite," *Nucl. Sci. Eng.*, 15, 69 (1963).

I15. J. P. Moore, R. S. Graves, and D. L. McElroy, "Thermal and
      Electrical Conductivities of Unirradiated and Irradiated
      Graphite from 300 to 1000 K," *Trans. Am. Nucl. Soc.* (U.S.),
      17, 145 (1973).

I16. S. Mrozowski, "The Nature of Artificial Carbons," *Proc. Soc.
      Chem. Ind.*, 7 (1958).

I17. A. I. Pobozhikhin, V. I. Volga, A. S. Kotosonov, I. F.
      Nikol'skaya, and A. V. Demin, "Electrophysical Properties of
      Artificial Graphites," *Izv. Akad. Nauk. SSSR, Ser. Neorg.
      Mater.*, 12(7), 1310 (1976).

I18. R. E. Taylor and W. D. Kimbrough, "Thermophysical Properties
      of ATJS Graphite at High Temperatures," *Carbon*, 8, 665 (1970).

I19. P. Wagner, J. A. O'Rourke, and P. E. Armstrong, "Porosity
      Effects in Polycrystalline Graphite," *J. Am. Ceram. Soc.·
      (U.S.), 55, 214 (1972).

J. Nongraphitizable Carbons[*]

J1. K. Antonowicz, L. Cacha, and J. Turlo, "Switching Phenomena
     in Glassy Carbon," *Carbon*, 11, 1 (1973).

J2. W. Bücker, "ESR, Seebeck, and Hall Measurements on an Amorphous
     System with Hopping Conduction," *J. Non Cryst. Solids*, 18, 11
     (1975).

_____

[*]See also A10, V12, W32, W33.

J3. W. Bücker, "Preparation and DC Conductivity of an Amorphous
     Organic Semiconducting System," *J. Non Cryst. Solids,* 12, 115
     (1973).

J4. L. R. Bunnell, "Structural Development in Vitreous Carbons as
     Observed by Scanning Electron Microscopy," *Carbon,* 12, 693
     (1974).

J5. A. V. Dmitriev and S. V. Shulepov, "Temperature Dependence of
     Specific Resistivity and Hall Coefficient in Non-Graphitizing
     Materials," *Vopr. Fiz. Tverd. Tela,* 7, 57 (1977).

J6. D. R. Hunt, G. M. Jenkins, and T. Takazawa, "The Effect of
     Tensile Stress upon the Resistivity of a Polymeric Carbon,"
     Carbon, 14, 105 (1976).

J7. G. M. Jenkins and K. Kawamura, "Structure of Glassy Carbon,"
     Nature, 231, 175 (1971).

J8. K. Kawamura and T. Tsuzuku, "Partial Graphitization in Relation
     to the Porosity of Glassy Carbons," *Carbon,* 12, 352 (1974).

J9. R. R. Saxena and R. H. Bragg, "Electrical Conduction in Glassy
     Carbon," *J. Non Cryst. Solids,* 28, 45 (1978).

J10. R. R. Saxena and R. H. Bragg, "Negative Magnetoresistance in
     Glassy Carbon," *Phil. Mag.,* 36, 1445 (1977).

J11. T. Shimada and T. Kikuchi, "Neutron Irradiation Effects of
     Electrical Properties in Glassy Carbon," *J. Phys. Soc. Japan,*
     20(7), 1288 (1965).

J12. T. Tsuzuku and K. Saito, "Electric Induction in Glassy Carbons,"
     Jpn. J. Appl. Phys., 5(8), 738 (1966).

J13. T. Yamaguchi, "Galvanomagnetic Properties of Glassy Carbon,"
     Carbon, 1, 47 (1963).

J14. T. Yamaguchi, "Thermoelectric Power of Glassy Carbon at High
     Temperature--Letter to the Editor," *Carbon,* 1, 535 (1964).

J15. S. Yamada and H. Sato, "Some Properties of Glassy Carbon,"
     J. Chem. Soc. Japan (Indust. Chem. Sect.), 65, 1139 (1962).

K. Fibers[*]

K1. R. Bacon, "Growth, Structure and Properties of Graphite Whisk-
     ers," *J. Appl. Phys.,* 31, 283 (1960).

K2. R. Bacon, "Carbon Fibers from Rayon Precursors," in *Chemistry
     and Physics of Carbon,* Vol. 9 (P. L. Walker, Jr. and P. A.
     Thrower, eds.), Marcel Dekker, New York, 1973.

[*]See also A10.

K3. C. Bazan, "The Electrical Resistivity of Carbon Filaments,"
     Acta. Phys. Polon., 423 (1966).

K4. A. A. Bright, "Negative Magnetoresistance in Pregraphitic
     Carbons," *Carbon*, 17, 255 (1979).

K5. A. A. Bright and L. S. Singer, "The Electronic and Structural
     Characteristics of Carbon Fibers from Mesophase Pitch," *Carbon*,
     17(1), 1978.

K6. D. Crawford and D. J. Johnson, "High Resolution Electron
     Microscopy of High-Modulus Carbon Fibres," *J. Microsc.*, 94,
     51 (1971).

K7. B. L. Elphick and R. Batty, "The Resistance of Carbon Filaments
     within the Frequency Range 0.1 to 4.0 GHz," *AWRE Rep.* (1977).

K8. H. M. Ezekiel, "Electrical Resistivity and Young's Modulus of
     Graphite Fibers," *J. Appl. Phys.*, 41, 5351 (1970).

K9. C. Herinckx, R. Perret, and W. Ruland, "Interstitial Compounds
     of Potassium with Carbon Fibers," *Carbon*, 10, 711 (1972).

K10. T. C. Holzschuh and W. J. Gajda, Jr., "DC Electrical Behavior
     of Graphite Fibers," *IEEE*, XI, 394 (1977).

K11. T. Koyama and M. Endo, "Electrical Resistivity of Carbon Fibers
     prepared from Benzene," *Jpn. J. Appl. Phys.*, 13, 1175 (1974).

K12. R. Perret and W. Ruland, "The Microstructure of PAN-base Carbon
     Fibres," *J. Appl. Crystallogr.*, 3, 525 (1970).

K13. N. B. Pokrovskaya et al., "Electrically conducting Fibers with
     Carbon Black Filler," *Fibre Chem.*, 8, 391 (1976).

K14. N. B. Pokrovskaya, N. N. Dolotova, A. A. Nikitin, and V. I.
     Maiboroda, "Conducting Carbon Black-reinforced Fibers," *Khim.
     Volokna*, 4, 39 (1976).

K15. W. N. Reynolds, "Structural and Physical Properties of Carbon
     Fibers," in *Chemistry and Physics of Carbon*, Vol. 11 (P. L.
     Walker, Jr. and P. A. Thrower, eds.), Marcel Dekker, New York,
     1973).

K16. D. Robson, F. Y. I. Assabghy, and D. J. E. Ingram, "Some Elec-
     tronic Properties of Polyacrylonitrile-based Carbon Fibers,"
     J. Phys. (U.K.), D5, 169 (1972).

K17. D. Robson, F. Y. I. Assabghy, E. G. Cooper, and D. J. E. Ingram,
     "Electronic Properties of High-Temperature Carbon Fibers and
     Their Correlations," *J. Phys.* (U.K.), D6, 1822 (1973).

K18. D. Robson, F. Y. I. Assabghy, D. J. E. Ingram, "An Electron
     Spin Resonance Study of Carbon Fibres on Polyacrylonitrile,"
     J. Phys. (U.K.), D4, 1426 (1971).

K19. F. L. Vogel, "Changes in Electrical Resistivity and Mechanical
     Properties of Graphite Fibers after Nitration," *Carbon*, 14,
     175 (1976).

K20. F. L. Vogel and R. Popowich, "Changes of Electrical Resistivity
      of Graphite Fibers with Nitration," *ACS Symp.*, *Ser. 21 (Pet.
      Deriv. Carbons, Symp., 1975), 411 (1976).*

K21. B. J. Wicks, "Microstructural Disorder and the Mechanical Prop-
      erties of Carbon Fibers," *J. Nucl. Mater.*, 56, 287 (1975).

K22. T. Yamaguchi, "Electronic Properties of Carbonized Polyacryl-
      onitrile Fibers--Letter to the Editor," *Carbon*, 2, 95 (1964).

K23. N. N. Yermolenko and R. N. Sevasternko, "Investigation of
      Certain Electrophysical Properties of Modified Carbon Fibres,"
      Izv. Akad. Nauk. Uzb. SSR, Ser. Neorg. Mater., 9(11), 1920-
      1924 (1973).

L. Carbon Films[*]

L1.   C. J. Adkins, S. M. Freake, and E. M. Hamilton, "Electrical
      Conduction in Amorphous Carbons," *Phil. Mag.*, 22, 183 (1970).

L2.   S. Aisenberg and R. Chabot, "Ion-Beam Deposition of Thin Films
      of Diamondlike Carbon," *J. Appl. Phys.*, 42, 2953 (1971).

L3.   D. A. Anderson, "Electrical and Optical Properties of Amorphous
      Carbon Prepared by the Glow Discharge Technique," *Phil. Mag.*,
      35, 17 (1977).

L4.   K. Antonowicz, "Possible Superconductivity at Room Temperature,"
      Nature, 247, 358 (1974).

L5.   K. Antonowicz, A. Jesmanowicz, and J. Wieczorek, "Switching
      Phenomena in Amorphous Carbon," *Carbon*, 10, 81 (1972).

L6.   A. Devenyi, et al., "Electrical Transport and Structure of
      Vacuum Deposited Carbon Films," *Proceedings International
      Conference of Physics and Chemistry of Semiconductor Helero-
      junctions and Layer Structures, October 1970,* Akad. Kiado,
      Budapest, 1971, pp. 97-104.

L7.   B. V. Derjaguin and D. V. Fedoseev, "Physico-Chemical Synthesis
      of Diamond in Metastable Range," *Carbon*, 11, 299 (1973).

L8.   B. V. Derjaguin and D. B. Fedoseev, "The Synthesis of Diamond
      at Low Pressure--Diamond Crystals are Grown in a Gas Rich in
      Carbon," *Sci. Amer.*, 233, 102 (1975).

L9.   J. N. Fox, "Low Temperature Characteristics of Carbon Films,"
      Cyrogenics, 12, 438 (1972).

L10. J. J. Hauser, "Hopping Conductivity in Amorphous Carbon Films,"
      Solid State Comm., 17, 1577 (1975).

---

[*]See also V1, V2, V6, V9-V11, V13, V16-V18, V20, V21, W9, W36.

L11. K. Kakinoki, "Deposited Carbon Film Resistors--The Key to the Development of Miniature Incombustible Resistors," *JEE*, 76, 65 (March 1973).

L12. D. S. Kupperman, C. S. Chau, and H. Weinstock, "Electrical Resistivity of Carbon Films," *Carbon*, 11, 171 (1973).

L13. M. L. A. MacVicar, "Amorphous Carbon Films: Conduction Across Metal/Carbon/Metal Sandwiches," *J. Appl. Phys.*, 41, 4765 (1970).

L14. I. S. McLintock and J. C. Orr, "Evaporated Carbon Films," in *Chemistry and Physics of Carbon*, Vol. 11 (P. L. Walker, Jr. and P. A. Thrower, eds.), Marcel Dekker, New York, 1973.

L15. I. S. McLintock and J. C. Orr, "Comments on the paper 'Electrical Resistivity of Carbon Films' by D. S. Kupperman, et al.," *Carbon*, 12, 91 (1974).

L16. M. Morgan, "Electrical Conduction in Amorphous Carbon Films," *Thin Solid Films*, 7(5), 313 (1971).

L17. A. Oberlin, "Study of Thin Amorphous and Crystalline Carbon Films by Electron Microscopy," *Phil. Mag.*, 32, 833 (1975).

L18. E. G. Spencer, P. H. Schmidt, D. C. Joy, F. J. Sansalone, "Ion-Beam Deposited Polycrystalline Diamond-like Films," *Appl. Phys. Lett.*, 29, 118 (1976).

M. Doped Carbons and Graphites[*]

M1. A. A. Belyi and A. Ovvchinnikov, "Electronic Structure of Doped Carbon," *Theor. Exp. Chem.*, 11, 599 (1976).

M2. J. Bulawa, S. Mrozowski and H. S. Vagh, "Dependence of the Hall Coefficient of Soft Carbons on Heat-Treatment and Doping," *Carbon*, 10, 207 (1972).

M3. P. Delhaes, "Proprietes Electroniques d'un Pyrocarbone Dope au Bore," *Comp. Ren.* (France), 261, 1298 (1965).

M4. P. Delhaes and A. Marchand, "Proprietes Electroniques d'un Coke de Brai Dope au Bore," *Carbon*, 3, 115 (1965).

M5. P. Delhaes and A. Marchand, "Proprietes Electroniques d'un Graphite Polycristallin Dope au Bore," *Carbon*, 3, 125 (1965).

M6. T. Kimura and K. Yazawa, "Effect of Chemical Impurities on the Galvanomagnetic Properties of Carbon," *Tanso*, 83, 139 (1975).

M7. A. S. Kotosonov, V. A. Vinniko, A. J. Polozhichin, and V. P. Sosedov, "Effect of chlorine on the Changes of Electronic Properties in the Course of Graphitization," *Carbon*, 8, 389 (1970).

---

[*]See also G15, G16, N8, N15, N16.

M8.   A. I. Lutcov, V. I. Volga, B. K. Dymov, V. N. Mikhailov, A. S.
      Tarabanov, and V. N. Bobkovskii, "Investigation of the Thermal
      and Electrical Conductivities of Siliconated Graphite," *Teplofiz
      Vys. Temp.* (USSR), 10, 1002 (1972) [trans. *High Temp.* (U.S.),
      10, 901 (1972)].

M9.   A. Marchand, "Electronic Properties of Doped Carbons," *Chemistry
      and Physics of Carbon,* Vol. 7 (P. L. Walker, Jr., ed.), Marcel
      Dekker, New York, 1971).

M10. A. Marchand and E. Dupart, "Proprietes Electronique d'un Pyro-
      carbon Dope avec du Bore: Evolution en Fonction du Taux de
      Bore," *Carbon,* 5, 453 (1967).

M11. A Marchand and J. V. Zanchetta, "Proprietes Electroniques d'un
      Carbone Dope a l'Azote," *Carbon,* 3, 483 (1966).

M12. S. Mrozowski, "Electrical Resistivity in Interstitial Compounds
      of Graphite," *J. Chem. Phys.*, 21, 492 (1953).

M13. D. E. Soule, "Change in Fermi Surfaces of Graphite by Dilute
      Acceptor Doping," *IBM J. Res. Dev.*, 268, July 1964.

M14. D. E. Soule, "The Effect of Boron on the Electronic Properties
      of Graphite," *Proc. 5th Conf. on Carbon,* 2, 13 (1961).

M15. P. V. Tamarin, A. B. Batdalov, V. I. Volga, "The Effect of
      Doping on Some Electrical Properties of Graphite," *Soviet Phys.
      Solid State,* 13, 2368 (1972).

M16. P. Wagner and J. M. Dickinson, "Ambient and High Temperature
      Experiments on Boron Doped Polycrystalline Graphites," *Carbon,*
      8, 313 (1970).

M17. S. Yajima and T. Hirai, "Siliconated Pyrolytic Graphite; Part 1,
      Preparation and Some Properties; Part 2, The State of Silicon
      Present in Siliconated Pyrolytic Graphite," *J. Mater. Sci.,* 4,
      416/424 (1969).

M18. S. Yajima and T. Hirai, "Siliconated Pyrolytic Graphite, Part 3,
      Structural Features," *Sci. Rep. Res. Instit. Tohoku Univer.,*
      21, A, 208-9 (1969).

M19. S. Yajima and T. Hirai, "Siliconated Pyrolytic Graphite: Part
      3, Structural Features; Siliconated Pyrolytic Graphite; Part 4,
      Electrical Resistivity," *J. Mater. Sci.,* 4, 685 (1969).

M20. F. I. Zorin, A. S. Kotosonav, and V. I. Volga, "Study of Nega-
      tive Magnetoresistance of Boron-doped Pyrolytic Carbons in the
      4.2-300 K Range," *Khim. Tverd. Topl.,* 6, 71 (1976).

N. Neutron and Electron--Irradiated Carbons and Graphites[*]

N1. S. B. Austerman and J. E. Hove, "Irradiation of Graphite at Liquid Helium Temperatures," *Phys. Rev.*, 100, 1214 (1955).

N2. A. Ammar and D. A. Young, "Electrical Conduction in Irradiated Pyrolytic Graphite," *Brit. J. Appl. Phys.*, 15, 131 (1964).

N3. S. Aranson, D. G. Schweitzer, R. M. Singer, and J. G. Davis, "Electrical Properties of Irradiated and Periodically Annealed Graphite (Letter to the Editor)," *J. Nucl. Mater.*, 12, 257 (1964).

N4. L. C. Blackman, G. Saunders, and A. R. Ubbelhode, "Radiation Damage in Well Oriented Pyrolytic Graphite," *Proc. Phys. Soc. (London)*, 78, 1048 (1961).

N5. L. Bochirol and E. Bonjour, "The Irradiation at Low Temperature by Neutrons and Electrons--The Measurement of Stored Energy and Variations of Electrical Resistivity," *Carbon*, 6, 742 (1968).

N6. J. D. Cooper, J. P. Smith, J. Woore, and D. A. Young, "Shubnikov-de Haas Oscillations in Neutron-irradiated Graphite," *J. Phys. Chem., Proc. Phys. Soc.*, 4, 442 (1971).

N7. R. O. Dillon, I. L. Spain, and J. W. McClure, "Galvanomagnetic Effects in Graphite III: The Effects of Neutron Irradiation on the Quantum-Limit Shubnikov-de Haas Oscillations," *J. Phys. Chem. Solids*, 39, 1071 (1978).

N8. M. Eto and T. Oku, "Change in Electrical Resistivity of Nuclear Graphite during Compression Tests and a Model for its Deformation and Fracture Mechanism," *J. Nucl. Mater.* (Netherlands), 54, 245 (1974).

N9. Y. Hishiyama, S. Mrozowski, and H. S. Vagh, "Studies of Negative Magnetoresistance in Neutron Irradiated Polycrystalline Graphite and in Boronated Kish Graphite," *Carbon*, 9, 367 (1971).

N10. T. Iwata, T. Nihira, and T. Ohmichi, "Recovery of the Basal Plane Electrical Resistivity in Electron-Irradiated Graphite," *Carbon*, 6, 742 (1968).

N11. T. Iwata, T. Nihira, and H. Matsuo, "Low Temperature Electron-Irradiation Damage and Recovery in Pyrolytic Graphite," *Phys. Soc. Japan*, 33, 1060 (1972).

N12. K. Kawamura and T. Tsuzuku, "A Study on Recovery Kinetics of Radiation Damage in Graphite," *J. Nucl. Mater.* (Netherlands), 60, 43 (1976).

[*]See also I11, I12, I15, J11, O2, P10, P15, P16, P23, Q57, W39.

N13. M. W. Lucas and E. W. J. Mitchell, "The Threshold Curve for the Displacement of Atoms in Graphite: Experiments on the Resistivity Changes Produced in Single Crystals by Fast Electron Irradiation at 15 K," *Carbon*, 1 (1964).

N14. H. Matsuo, "Thermal Annealing Effects on the Electrical Resistivity of Reactor Grade Graphite Irradiated with Neutrons at 250°C and 350°C," *J. Nucl. Mater.* (Netherlands), 41, 235 (1971).

N15. H. Matsuo and T. Honda, "Annealing of Reactor Grade Graphite Irradiated with Neutrons at 350°C--Effect on Electrical Resistivity and Thermal Conductivity," *J. Nucl. Mater.* (Netherlands), 45, 79 (1972).

N16. S. Mrozowski, "Electron Spin Resonance in Neutron Irradiated and in Doped Polycrystalline Graphite--Part I," *Carbon*, 3, 305 (1965).

N17. S. Mrozowski, "Electron Spin Resonance in Neutron Irradiated and in Doped Polycrystalline Graphite--Part II," *Carbon*, 4, 227 (1966).

N18. H. L. Pitner, "Irradiation Behavior of Pocographites," *Carbon*, 9, 637 (1971).

N19. A. I. Polozhikhin, Yu S. Virgil'ev, A. S. Kotosonov, G. F. Efremova, and I. P. Kalyagina, "Effect of the Degree of Graphitization of a Carbon Material on the Change in Magnetoresistive Effects During Neutron Irradiation," *At. Energ.* (USSR), 35, 207 (1973) [trans. *Soviet At. Energ.*, 855].

N20. R. L. Powell and P. Wagner, "Irradiation Effects on Low Temperature Thermal and Electrical Conductivities of Two Graphites," *Carbon*, 8, 690 (1970).

N21. J. Rappenau, G. Micaud, A. Pacault, A. Marchand, and J. Amiell, "Evolution of Electronic Properties of Carbons with Neutron Irradiation," *Carbon*, 8, 55 (1970).

N22. J. Rappenau, G. Micaud, A. Marchand, A. Pacault, and J. Amiell, "Evolution of the Different Classes of Carbons by Neutron Irradiation," *Carbon*, 14, 53 (1976).

N23. W. N. Reynolds, "Radiation Damage in Graphite," in *Chemistry and Physics of Carbon*, Vol. 2 (P. L. Walker, Jr., ed.), Marcel Dekker, New York, 1966, pp. 121-196.

N24. W. N. Reynolds and P. R. Goggin, "The Annealing of Electron Irradiation Damage in Graphite," *Phil. Mag.*, 5, 1049 (1960).

N25. T. Tsuzuku and S. Arai, "Annealing Effects on Galvanomagnetic Properties of Heavily Radiation-damaged Graphite," *Jpn. J. Appl. Phys.*, 10, 580 (1971).

N26. Yu. S. Virgil'ev and V. G. Makarenko, "Electrical Resistance of Carbon Materials and Its Alteration by Irradiation," *Izv. Akad. Nauk SSSR, Inorg. Mater.*, 9, 1546 (1973). [trans. *Inorg. Mater.* (U.S.), 9, 1375 (1973)].

N27. S. Yugo, "ESR Studies of Neutron-Irradiated Graphite. II.
      Investigation of Electronic Structure," *Rep. Univ. Electro-
      Commun.* (Japan), 27, 303 (1977).

0. C-Axis Conductivity*

01. A. S. Bender and D. A. Young, "The Anisotropy of Carrier Life-
     time in Graphite," *J. Phys.* (U.K.), C, 6, 43 (1973).

02. T. Iwata, T. Nihira, and H. Matsuo, "Irradiation and Annealing
     Effects on the C-Axis Electrical Resistivity of Graphite," *J.
     Phys. Soc. Japan,* 36, 123 (1974).

03. K. Kawamura, Y. Ouchi, H. Oshima, and T. Tsuzuku, "Electrical
     Conduction in C-Direction of Highly Crystalline Graphites in
     Relation to Structural Perfection," *J. Phys. Soc. Japan,* 46,
     587 (1979).

04. N. Okuyama, "Longitudinal Magnetoresistance Along the C-Axis
     of Graphite in the Quantum Limit," *Rep. Univ₀ Electro-Commun.*
     (Japan), 28, 11 (1977).

05. N. Okuyama, H. Yasunaga, S. Minomora, and K. Takeya, "Dependence
     of the Resistance on Pressure in the C-Direction of Pyrolytic
     and Natural Graphite," *Jpn. J. Appl. Phys.,* 10, 1645 (1975).

06. S. Ono, "C-Axis Resistivity of Graphite in Relation to Stacking
     Faults," *J. Phys. Soc. Japan,* 40, 498 (1976).

07. W. Primak, "C-Axis Electrical Conductivity of Graphite," *Phys.
     Rev₀,* 103, 544 (1956).

08. Yu A. Pospelov, "On the Dependence of σ_{zz} of Graphite on
     Temperature and Pressure," *Fiz₀ Tverd. Tela,* 6, 1525 (1964)
     [trans. *Sov₀ Phys. Solid State,* 6, 1193 (1964)].

09. A. R. Saha, A. K. Das, and P. K. Banerjee, "Microwave Studies
     of C-Axis Resistivity of Pyrolytic Graphite," *Indian J. Phys.,*
     46, 12, 537-46 (1972).

010. I. L. Spain, "C-Axis Conduction in Graphite," *The Physics of
     Semimetals and Narrow Gap Semiconductors* (Carter and Bate, eds.),
     Pergamon Press, Oxford, 1971, pp. 177-185.

011. I. L. Spain and J. A. Woollam, "The Longitudinal Magnetoresis-
     tance of Graphite $\rho_{zz}(H_z)$ in High Magnetic Fields," *Solid State
     Commun.,* 9(18), 1581 (1971).

012. D. Z. Tsang and M. S. Dresselhaus, "The C-Axis Electrical Con-
     ductivity of Kish Graphite," *Carbon,* 14, 43 (1976).

—————————
*See also E6, E12, F7, F8, G3, G11-G13, G14, Q79, S7, S13.

013. A. R. Ubbelohde, "The Anisotropy of Graphite," *Endeavour*, 24, 92, 63 (1965).

014. G. Wagoner, "Spin Resonance of Charge Carriers in Graphite," *Phys. Rev.*, 118, 647 (1960).

015. D. A. Young, "Evidence for State Localization in Pyrolytic Graphite," *Phys. Status Solidi*, B, 43, K143 (1971).

016. D. A. Young, "On the C-Axis Conductivity of Pyrolytic Graphite," *Phys. Status Solidi*, B, 50, K143 (1972).

P. Diamond[*]

P1. I. G. Austin and R. Wolfe, "Electrical and Optical Properties of a Semiconducting Diamond," *Proc. Phys. Soc.* (London), B, 69, 329 (1956).

P2. R. T. Bate and R. K. Willardson, "Hall Coefficient and Magneto-resistance of Semiconducting Diamond," *Proc. Phys. Soc.* (London), 74, 363 (1959).

P3. A. V. Bogdanov and V. A. Presnov, "Voltage Current Characteristics of Synthetic Semiconductor Diamonds, Doped with Boron During Synthesis," *Izv. Vuz. Fiz.* (USSR), 9, 7 (1978) [trans. *Sov. Phys. J.* (U.S.)].

P4. J. J. Brophy, "Preliminary Study of the Electrical Properties of a Semiconducting Diamond," *Phys. Rev.*, 99, 1336 (1955).

P5. F. P. Bundy, H. M. Strong, and R. H. Wentorf, Jr., "Methods and Mechanisms of Synthetic Diamond Growth," in *Chemistry and Physics of Carbon*, Vol. 10 (P. L. Walker, Jr. and P. A. Thrower, eds.), Marcel Dekker, New York, 1973.

P6. V. P. Butuzov, V. A. Laptev, V. P. Dunin, B. I. Zadneprovskii, and N. G. Sanzharlinskii, "Growth of Diamond Whiskers in a Metal-Carbon System at High Temperatures and Pressures," *Dokl. Akad. Nauk. SSSR*, 225, 88 (1975).

P7. V. P. Butuzov, V. A. Laptev, V. A. Presnov, and Yu. M. Rotner, "Preparation and Investigation of Synthetic Semiconducting Diamonds of Various Conductivity Types," *Dokl. Akad. Nauk SSSR*, 226, 328 (1976).

P8. J. F. H. Custers, "Semiconductivity of a Type II b Diamond," *Nature* (London), 176, 173 (1955).

P9. J. J. Hauser and J. R. Patel, "Hopping Conductivity in C-Implanted Amorphous Diamond or How to Ruin a Perfectly Good Diamond," *Solid State Comm.* (U.S.), 18, 789 (1976).

[*]See also C11, C13, C26, C27, L2, L7, L8, L18.

P10. S. M. Horszowski, "The Charge Carrier Mobility of Electron-Irradiated Natural Semiconducting Diamond," *Phil. Mag.*, 34(6), 1109 (1976).

P11. E. Hulicius, "Diamonds in Electronics," *Cesk. Cas. Fis.*, A (Czechoslovakia), 26, 414 (1976).

P12. P. J. Kemmy and P. T. Wedepohl, "Semiconducting Diamond," in *Physical Properties of Diamond* (R. Berman, ed.), Clarendon Press, Oxford, 1956.

P13. V. A. Laptev, V. A. Presnov, and Yu M. Rotner, "Preparation and Study of Synthetic Semiconductor Diamonds with a Different Type of Conductivity," *Dokl. Akad. Nauk SSR,* 226, 378 (1976).

P14. E. C. Lightowlers and A. T. Collins, "Electrical Transport Measurements on Synthetic Semiconducting Diamond," *Phys. Rev.*, 151, 685 (1966).

P15. V. G. Malogolovets and A. G. Gontar, "Optical and Electrical Properties of Synthetic Diamond Single Crystals, Irradiated by Neutrons," *Ukr. Fiz. Zh.* (USSR), 23, 860 (1978).

P16. B. Massarani and J. C. Bourgoin, "Defects at Low Temperatures in Electron-Irradiated Diamond," *Phys. Rev.*, B, 14, 3682 (1976).

P17. B. Massarani, J. C. Bourgoin, and R. M. Chrenko, "Hopping Conduction in Semiconducting Diamond," *Phys. Rev.*, B, 17, 1758 (1978).

P18. R. E. Mutch and R. A. Raal, "Electrical Resistivity Changes Observed in a Semiconducting Diamond After Heat Treatment," *Nature* (London), 184, 1857 (1959).

P19. I. A. Parfianovich, Yu. S. Mukhachev, and S. Yu. Borzenko, "Electrical Conductivity and Thermoelectric Power of Natural Diamonds," *Fiz. Tekh. Poluprovodn.* (USSR), 11, 1582 (1977).

P20. E. E. Semenova, G. A. Dubitskii, and V. N. Slesarev, "Synthesis of Semiconducting Diamond by a High-Pressure, High-Temperature Diffusion Method," *Pis'ma Zh. Tekh. Fiz.* (USSR), 4, 86 (1978).

P21. V. S. Tatarinov and Yu. S. Mukhachev, "Depolarization Currents in a Type-I Diamond During X-ray Excitation," *Lyumin. Spektr. Anal.,* 3, 126 (1974).

P22. Y. F. Tsag, K. P. Ananthanarayanan, P. J. Gielisse, and S. S. Mitra, "Electrical Conductivity of Heavily Doped Diamond," *J. Appl. Phys.*, 43, 3677 (1972).

P23. E. R. Vance, H. J. Millegee, and A. T. Collins, "The Effect of Heavy Neutron Irradiation on the Electrical Resistivity of Diamond," *J. Phys. D., J. Appl. Phys.*, 5, 140 (1972).

P24. V. S. Vavilov, "Semiconducting Diamond," *Phys. Status Sol.*, A (Germany), 31, 11 (1975).

P25. V. S. Vavilov, M. A. Gukasyan, M. I. Suseva, and E. A. Konorova,
      "Conductivity of Diamond Doped by Implantation of Phosphorus
      Ions," *Fiz. and Tekh. Poluprovodn* (USSR), 9, 1455 (1975) [trans.
      Sov. Phys.-Semicond. (U.S.), 9, 962 (1975)].

P26. V. S. Vavilov and E. A. Konorova, "Semiconducting Diamonds,"
      Usp. Fiz. Nauk (USSR), 118, 611 (1976) [trans. *Sov. Phys. Usp.*
      (U.S.), 19, 301 (1976)].

P27. A. S. Vishnevskii and A. G. Gontar, "Electrical Conductivity
      of Synthetic Diamond Crystals," *Fiz. and Tekh. Poluprovodn.*
      (USSR), 11, 2024 (1977).

P28. P. T. Wedepohl, "Electrical and Optical Properties of Type II b
      Diamonds," *Proc. Phys. Soc.* (London), B, 70, 177 (1957).

P29. R. H. Wentorf, Jr. and B. P. Bovenkerf, "Preparation of Semi-
      conducting Diamonds," *J. Chem. Phys.*, 36, 1987 (1962).

P30. R. H. Wentorf, Jr. and K. A. Darrow, "Semiconducting Diamonds
      by Ion Bombardment," *Phys. Rev.*, A, 137, 1614 (1965).

Q. Intercalation Compounds of Graphite[*]

Q1. A. Aoki and S. Yajima, "High Temperature Resistivity of Pyro-
     lytic Graphite Bromine Residual Compounds," *J. Mater. Sci.*, 6,
     1338 (1971).

Q2. G. Bach and A. R. Ubbelohde, "Chemical and Electrical Behavior
     of Graphite-Metal Halide Compounds," *J. Chem. Soc.*, A, 23, 3699
     (1971).

Q3. F. Batallan, J. Bok, I. Rosenman, and J. Melin, "Electronic
     Structure of Intercalate Graphite by Magnetothermal Oscillations,"
     Phys. Rev. Lett., 41, 330 (1978).

Q4. A. S. Bender and D. A. Young, "Shubnikov-de Haas Oscillations
     in Graphite-Bromine," *Phys. Status Solidi*, B, 47, K95 (1971).

Q5. A. S. Bender and D. A. Young, "Fermi Surfaces in Graphite-
     Bromine, Studied by Shubnikov-de Haas Oscillations," *J. Phys.
     Chem., Solid State Phys.*, 5, 2163 (1972).

Q6. L. C. F. Blackman, J. F. Mathews, and A. R. Ubbelohde, "Elec-
     trical Properties of Crystal Compounds of Graphite. I. Con-
     ductance of Graphite/Bromine," *Proc. Roy. Soc.*, A, 256, 15
     (1060).

Q7. L. C. F. Blackman, J. F. Mathews, and A. R. Ubbelohde, "Elec-
     trical Properties of Crystal Compounds of Graphite. II. Acid
     salts of Graphite," *Proc. Roy. Soc.*, A, 258, 339 (1960).

---

[*]See also A3, A6, A21, G26, K9, K19, K20.

Q8. L. C. F. Blackman, J. F. Mathews, and A. R. Ubbelohde, "Electrical Properties of Crystal Compounds of Graphite. III. The Role of Electron Donors," *Proc. Roy. Soc.*, A, 258, 339 (1960).

Q9. J. Bok, "Quantum Magnetothermal Oscillations of Intercalation Compounds of Graphite," *Proceedings of the International Conference High Magnetic Fields,* Oxford, 1978.

Q10. D. D. L. Chung, "Structural Studies of Graphite Intercalation Compounds," *J. Electr. Mater.*, 7, 189 (1978).

Q11. D. D. L. Chung and M. S. Dresselhaus, "Magnetoreflection Study of Graphite Intercalated with Bromine," *Solid State Comm.* (U.S.), 9, 227 (1976).

Q12. D. D. L. Chung and M. S. Dresselhaus, "Magneto-Optical Studies of Graphite Intercalation Compounds," *Physica*, B, 89, 131 (1977).

Q13. G. Colin and J. F. Boissoneau, "Evolution of Resistivity Between 90 and 298 K and Thermoelectric Power of Bromine Saturated Pyrocarbon," *J. Solid State Chem.*, 5, 342-5 (1972).

Q14. G. Colin and E. Durizot, "Intercalation Compounds of $CdCl_2$ in Graphite: Formation, Structural Data, and Electrical Resistivity," *J. Mater. Sci.*, 9, 1994 (1974).

Q15. G. Colin and C. Mazieres, "Mesure de la Resistivite par Courant Induits. II. Problème Posé Par 1-Anisotropie Magnetique. III. Etude in situ de I'interaction Brome Pyrocarbone," *J. Chim. Phys. Physico-Chim. Biol.*, 67, 323 (1970).

Q16. M. Crespin, D. Tchoubar, L. Gatineau, F. Beguin, R. Setton, "Influence of Intercalation-Desorption on the Mutual Arrangement of the Elementary Sheets in a Partially Graphitized Carbon," *Carbon*, 15, 303 (1977).

Q17. P. Delhaes, "Physical Properties of Graphite Lamellar Compounds with Alkali Metals and Halogens," *Mater. Sci. Eng.*, 31, 225 (1977).

Q18. G. Dresselhaus and M. S. Dresselhaus, "Intercalation Ionization in Graphite-Halogen Compounds," *Mater. Sci. Eng.*, 31, 235 (1977).

Q19. M. S. Dresselhaus, G. Dresselhaus, and J. E. Fischer, "Graphite Intercalation Compounds: Electronic Properties in the Dilute Limit," *Phys. Rev.*, B, 15, 3180 (1977).

Q20. W. D. Ellenson and D. Semmingsen, "Neutron Scattering Studies of Alkali Metal-Graphite Intercalation Compounds," *Mater. Sci. Eng.*, 31, 137 (1977).

Q21. E. L. Evans and J. M. Thomas, "Ultra Microstructural Characteristics of Some Intercalates of Graphite: An Electron Microscope Study," *J. Solid St. Chem.*, 14, 99 (1975).

Q22. J. E. Fischer, T. E. Thompson, G. M. T. Foley, D. Guerard, M. Hoke, and F. L. Lederman, "Optical and Electrical Properties of Graphite Intercalated with HNO_3," *Phys. Rev. Lett.* (U.S.), 37, 779 (1976).

Q23. J. E. Fischer, "A Simple Model of the a-Axis Conductivity in Graphite Intercalation Compounds," *Carbon*, 15, 161 (1977).

Q24. J. E. Fischer, "Electronic Properties of Graphite Intercalation Compounds, Intercalation Compounds of Graphite Proceedings," *Mater. Sci. Eng.*, 31, 211 (1977).

Q25. J. E. Fisher, "Electronic Properties of Graphite Intercalation Compounds," in *Physics and Chemistry of Materials with Layered Structures*, Vol. 5: *Intercalation Compounds* (F. Levy, ed.), D. Reidel, Dordrecht, Holland, 1970.

Q26. J. E. Fischer and T. E. Thompson, "Graphite Intercalation Compounds," *Physics Today*, 7, 36 (1978).

Q27. J. E. Fischer, T. E. Thompson, and F. L. Vogel, "Free Carrier Plasma in Graphite Compounds," *ACS Symp. Ser.*, 21 (Pet. Deriv. Carbons, Symp., 1975), 418 (1976).

Q28. G. M. T. Foley, C. Zeller, E. R. Falardeau, and F. L. Vogel, "Room Temperature Electrical Conductivity of a Two-Deminesional Synthetic Metal: AsFe₅ Graphite," *Solid State Comm.*, 24, 371 (1977).

Q29. H. Fuzellier, J. Melvin, and A. Herold, "Electrical Conductivity of Lamellar Compounds of Graphite-SbF₅ and SbCl₅ (in French)," *Carbon*, 15, 45 (1977).

Q30. L. A. Girifalco and T. O. Montelbano, "Preparation and Properties of Barium-Graphite Compound," *J. Mater. Sci.*, (U.K.), 11, 1036 (1976).

Q31. D. Guerard, G. M. T. Foley, M. Zanini, and J. E. Fischer, "Electronic Structure of Donor-Type Graphite Intercalation Compounds," *Nuovo Cimento,* B (Italy), B, 38, 410 (1977).

Q32. D. Guerard, G. M. T. Foley, M. Zanini, and J. E. Fischer, "Electronic Structure of Donor-Type Graphite Intercalation Compounds," *Nuovo Cimento,* B, 38, 410 (1977).

Q33. S. G. Hegde, E. Lerner, and J. G. Daunt, "Thermal and Electrical Conductivities of Exfoliated Graphite at Low Temperatures," *Cryogenics* (U.K.), 13, 230 (1973).

Q34. G. Hennig, "The Properties of Interstitial Compounds of Graphite I. The Electronic Structure of Graphite Bisulphate," *J. Chem. Phys.* (U.S.), 19, 922 (1951).

Q35. G. Hennig and L. Meyer, "Search for Low Temperature Superconductivity in Graphite Compounds," *Phys. Rev.*, 87, 439 (1952).

Q36. Y. Hishiyama, A. Ono, M. Inagaki, and T. Tsuzuki, "Electronic Processes in Residue Compounds of Graphite Nitrate," *Jpn. J. Appl. Phys.*, 8, 1189 (1969).

Q37. Y. Hishiyama, A. Ono, M. Inagaki, and T. Tsuzuki, "Electronic Processes in Residue Compounds of Graphite Nitrate," *Jpn. J. Appl. Phys.*, 9(1), 159 (1970) [*Jpn. J. Appl. Phys.*, 8, 1189 (1969).]

Q38. Y. Hishiyama and A. Ono, "Galvanomagnetic Effects in Nitrate-Doped Graphite," *Jpn. J. Appl. Phys.*, 2(3), 265-67 (1970).

Q39. N. B. Hannay, T. H. Geballe, B. T. Matthias, K. Andres, P. Schmidt, and D. McNair, "Superconductivity in Graphitic Compounds," *Phys. Rev. Lett.*, 14, 225 (1965).

Q40. N. A. W. Holz and S. Rabii, "Energy Band Structure of Lithium-Graphite Intercalation Compound," *Mater. Sci. Eng.*, 31, 201 (1977).

Q41. M. Inagaki, J. C. Rouillan, G. Fing, and F. Delhaes, "Physical Properties of Graphite-Nitrate Residue Compound," *Carbon*, 15, 181 (1977).

Q42. T. Inoshita, K. Nakao, H. Kamimura, "Electronic Structure of Potassium-Graphite Intercalation Compound: C_8K," *J. Phys. Soc. Japan*, 43, 1237 (1977).

Q43. K. Kawamura, T. Saito and T. Tsuzuku, "Galvanomagnetic Properties of Graphite Intercalated with Nitrate," *Jpn. J. Appl. Phys.*, 17, 1207 (1978).

Q44. K. Kawamura, T. Saito, and T. Tsuzuku, "Phase Transitions of Graphite Nitrate Residue Compounds in Galvano-Magnetic Properties," *Carbon*, 13, 452 (1975).

Q45. S. K. Khanna, E. R. Falardeau, A. J. Heeger, and J. E. Fischer, "Conduction Electron Spin Resonance in Acceptor-Type Graphite Intercalation Compounds," *Solid State Comm.*, 25, 1059 (1978).

Q46. E. A. Kmetko, "Electronic Properties of Carbons and Their Interstitial Bisulphate Compounds," *J. Chem. Phys.* (U.S.), 21, 2152 (1953).

Q47. Y. Koike, H. Suematsu, K. Higuchi, and S. Tanuma, "Superconductivity in Potassium Graphite Intercalation Compound C_8K," *Solid State Comm.*, 27, 623 (1978).

Q48. S. Loughin, R. Grayeski, and J. E. Fischer, "Charge Transfer in Graphite Nitrate and the Ionic Salt Model," *J. Chem. Phys.*, 69, 3740 (1978).

Q49. E. McKae and A. Herold, "Intercalation Compound Resistivity Measurements," *Mater. Sci. Eng.*, 31, 249 (1977).

Q50. K. Miyauchi, Y. Takahashi, and T. Mukaibo, "Anomalous Electric Resistivity Changes of Graphite-Bromine Residue Compound at High Temperature--Letters to the Editor," *Carbon*, 9, 807 (1971).

Q51. J. J. Murray and A. R. Ubbelohde, "Electronic Properties of Some Synthetic Metals Derived from Graphite," *Proc. Roy. Soc.* (London), A, 312, 371 (1969).

Q52. L. C. Olsen, S. E. Seeman, and H. W. Scott, "Expanded Pyrolytic Graphite: Structural and Transport Properties," *Carbon*, 8, 85-93 (1970).

Q53. D. G. Onn, G. M. T. Foley, and J. E. Fischer, "Resistivity
      Anomalies and Phase Transitions in Alkali-Metal Graphite Inter-
      calation Compounds," *Mater. Sci. Eng.*, 31, 271 (1977).

Q54. G. S. Parry, "Structural Ordering in Intercalation Compounds,"
      Mater. Sci. Eng., 31, 99 (1979).

Q55. D. A. Platts, D. D. L. Chung, and M. S. Dresselhaus, "Far-
      Infrared Magnetoreflection Studies of Graphite Intercalated
      with Bromine," *Phys. Rev.*, B, 15, 1087 (1977).

Q56. L. Pretronero, S. Stiassler, H. R. Zeller, and M. J. Rice,
      "Charge Distribution in c-Direction in Lamellar Graphite
      Acceptor Compounds," *Phys. Rev. Lett.*, 41, 763 (1978).

Q57. J. C. Rouillon and A. Marchand, "Composes Residuels de Pyro-
      carbone et de Brome: "Hysterese" du Diamagnetisme et de la
      Resistivite Par Rapport a la Température," *Comp. Rend.*, C,
      274, 112 (1972).

Q58. F. J. Salzano and M. Strongin, "Dimensionality of Superconduc-
      tivity in Graphite Lamellar Compounds," *Phys. Rev.*, 153, 533
      (1967).

Q59. E. J. Seykora, R. A. Klein, "An Organic Metal?," *Nature* (U.K.),
      248, 401 (1974).

Q60. A. W. Smith and N. S. Rasor, "Observed Dependence of Low-
      Temperature Thermal and Electrical Conductivity of Graphite on
      Temperature, Type, Neutron Irradiation and Bromination," *Phys.
      Rev.*, 104, 885 (1956).

Q61. I. L. Spain, A. R. Ubbelohde, and D. A. Young, "A Physico-
      chemical Technique for Evaluating Defects in Graphite," *J.
      Chem. Soc.* (London), 180, 920 (1964).

Q62. I. L. Spain and D. J. Nagel, "The Electronic Properties of
      Lamellar Compounds of Graphite—An Introduction," *Mater. Sci.
      Eng.*, 31, 183 (1977).

Q63. Y. Takahashi, H. Yamagata, and T. Mukaibo, "Structure and
      Properties of Graphite-Chromyl Chloride Lamellar Compounds,"
      Carbon, 11, 19 (1973).

Q64. T. E. Thompson, E. R. Falardeau, and L. R. Hanlon, "The Elec-
      trical Conductivity and Optical Reflectance of Graphite-SbF_5
      Compounds," *Carbon*, 15, 39 (1977).

Q65. J. M. Thomas, G. R. Millward, N. C. Davies, and E. H. Evans,
      "On 'Seeing' the Stacking Sequence in Graphite-$FeCl_3$ Inter-
      calates by High Resolution Electron Microscopy," *J. Chem. Soc.*
      (Trans.--Dalton), 1976, 2443 (1976).

Q66. A. R. Ubbelohde, "Electronic Properties of Graphite and Its
      Crystal Compounds in the Direction of the C-axis," *Proceedings
      5th Conference Carbon*, MacMillan, New York, 1962, p. 1.

Q67. A. R. Ubbelohde, "Properties of Synthetic Good Conductors of Electricity," *Nature*, 210, 404 (1966).

Q68. A. R. Ubbelohde, "Electrical Properties and Phase Transformations of Graphite Nitrates," *Proc. Roy. Soc.*, A, 304, 25 (1968).

Q69. A. R. Ubbelohde, "Charge Transfer Effects in Acid Salts of Graphite," *Proc. Roy. Soc.*, A, *Math. Phys. Sci.*, 309, 1498, 297 (1969).

Q70. A. R. Ubbelohde, "Anisotropy of Synthetic Metals," *Nature* (U.K.), 232, 43 (1971).

Q71. A. R. Ubbelohde, "Electronic Anomalies in Dilute Synthetic Metals," *Proc. Roy. Soc.* (London), A, 321, 445 (1971).

Q72. A. R. Ubbelohde, "Intercalation Overpotentials for Compound Formation by Graphite," *Carbon*, 10, 201 (1972).

Q73. A. R. Ubbelohde, "Electrical Anisotropy of Synthetic Metals Based on Graphite," *Proc. Roy. Soc.*, (London), A, 327, 289 (1972).

Q74. A. R. Ubbelohde, "Synthetic Metals: Comment," *Pontif. Acad. Sci.*, 2, 11 (1975).

Q75. A. R. Ubbelohde, "Carbons as a Route to Synthetic Metals," *Carbon*, 14, 1 (1976).

Q76. A. R. Ubbelohde, "Intercalation into Carbons with Varying Degree of Graphitization," *Mater. Sci. Eng.*, 31, 341 (1977).

Q77. A. R. Ubbelohde, L. C. F. Blackman, and J. F. Mathews, "Metallic Conduction in the Crystal Compounds of Graphite," *Nature*, 183, 454 (1959).

Q78. F. L. Vogel, "Electrical Resistivity of Nitrate-Intercalated Graphite Fibers," *Proceedings Fourth London International Conference Carbon and Graphite*, 1974, p. 332.

Q79. F. L. Vogel, Jr. and R'Sue Popawich, "Changes in the Electrical Resistivity and Structure of Graphite with Intercalation by Nitric Acid," *ACS Symp.*, (Petr. Der. Carbons Symp., 1975), 20(21), 461 (1975).

Q80. F. L. Vogel, "Changes in Electrical Resistivity and Mechanical Properties of Graphite Fibers after Nitration," *Carbon*, 14, 175 (1976).

Q81. F. L. Vogel, "The Electrical Conductivity of Graphite Intercalated with Superacid Fluorides: Experiments with Antimony Pentafluoride," *J. Mater. Sci.*, 12, 982 (1977).

Q82. F. L. Vogel, G. M. T. Foley, C. Zeller, E. R. Falardeau, and J. Gan, "High Electrical Conductivity in Graphite Intercalated with Acid Fluorides," *Mater. Sci. Eng.*, 31, 261 (1977).

Q83. S. Yajima, T. Hirai, and K. Aoki, "The Electrical Anisotropies of Pyrolytic Graphite and Its Compounds," *Rad. Eff.* (U.K.), 7, 55 (1970).

Q84. C. Zeller, G. M. T. Foley, E. R. Falardeau, and F. L. Vogel, "Measurement of Electrical Conductivity under Conditions of High Anisotropy in Graphite Intercalation Compounds," *Mater. Sci. Eng.*, 31, 255 (1977).

Q85. C. Zeller, L. A. Pendrys, and F. L. Vogel, "Electrical Transport Properties of Low-Stage AsF$_5$-Intercalated Graphite," (to be published).

R. Powders and Mixtures and Composites*

R1.   M. F. Boisard, "Etude Experimentale de la Variation Thermique du Coefficient de Hall de Carbones Pregraphitiques Pulverulents," *Compt. Rend.* (France), 258, 549 (1964).

R2.   P. Caillon, J. P. Reboul, and A. Toureille, "Physique des Solide-Conduction Electrique dans les Hauts-Polymeres Charges de noir de Carbone," *Compt. Rend.* (France), B, 272, 1074 (1971).

R3.   G. B. Demidarich and V. F. Kiselev, "Influence of Surface State of the Prismatic Faces of Dispersed Graphite on its Electric Conductivity Work Function and Thermoelectromotive Forces," *Zh. Fiz. Khim.* (USSR), 41, 684 (1967).

R4. · J. B. Donnet and A. Voet, *Carbon Black*, Marcel Dekker, New York, 1967.

R5.   E. O. Forster, "Electrical Conduction Mechanism in Carbon Filled Polymers," *IEEE Trans. Power Apparatus Systems* (PAS), 90, 913 (1971).

R6.   V. E. Gul, V. F. Blinov, M. G. Golubeva, and L. M. Ryabava, "Study of the Hall Effect in Carbon Black-filled Polyimide Films," *Plast. Massy*, 7, 55 (1976).

R7.   E. Kato, "Electrical Resistivity of Carbon Black-Kaolin Composite Sintered Materials I. Relation Between Resistivity and Carbon Content," *J. Chem. Soc. Japan*, 67, 2026 (1964).

R8.   E. Kato, "Electrical Resistivity of Carbon Black-Kaolin Composite Sintered Materials II. Effects of Firing Temperature on Resistivity," *J. Chem. Soc. Japan*, 68, 442 (1965).

R9.   E. Kato, "Electrical Resistivity of Carbon Black-Kaolin Composite Sintered Materials III. Inner Changes During Sintering and Their Effects on Resistivity," *J. Chem. Soc. Japan*, 68, 1038 (1965).

R10. E. Kato, "Electrical Resistivity of Carbons Black-Kaolin Composite Sintered Materials IV. Relations Between Temperature Dependence of Resistivity and Carbon Bulk Density," *J. Chem. Soc. Japan*, 69, 188 (1966).

---

*See also O5.

R11. E. Kato, "Electrical Resistivity of Carbon Black-Kaolin
      Composite Sintered Materials V. Effects of Firing Temperature
      on Resistivity-Temperature Dependence," *J. Chem. Soc. Japan*, 69,
      196 (1966).

R12. E. Kato and M. Hasegawa, "Electrical Resistivity of Carbon
      Black-Kaolin Composite Sintered Materials VI. Pressure
      Dependence of Resistivity of Carbon Black under Loose Contact,"
      J. Chem. Soc. Japan, 69, 384 (1966).

R13. E. Kato and M. Hasegawa, "Electrical Resistivity of Carbon
      Black-Kaolin Composite Sintered Materials VIII. Effects of
      Carbon Materials on Resistivity," *J. Chem. Soc. Japan* (Ind.
      Chem. Sec.), 69, 1117 (1966).

R14. E. Kato and M. Hasegawa, "Electrical Resistivity of Carbon
      Black-Kaolin Composite Sintered Materials IX. Effects of
      Firing Duration, Firing Atmosphere and Forming Process on
      Resistivity"; "X. Effects of Carbon Materials on Temperature
      Dependence of Resistivity," *J. Chem. Soc. Japan*, 69, 2108 (1966).

R15. E. Kato and M. Hasegawa, "Electrical Resistivity of Carbon
      Black-Kaolin Composite Sintered Materials XI. Resistivity-
      Temperature Dependence of the Compositions Containing Excess
      Silica," *J. Chem. Soc. Japan*, 70, 252 (1967).

R16. B. N. Klochko, E. E. Zachateiskii, et al., "Effect of the
      Degree of Dispersion and Structure of Carbon Black on its
      Electrical Resistance," *Kauch. Rezina*, 28, 22 (1969).

R17. S. Kubota, "Properties of Carbon Powders for Telephone Trans-
      mitter Having Strong Correlation to their Life," *Tanso*, 37,
      7 (1964).

R18. S. Kubata, "Studies on Carbon Powders for Telephone Transmitter-
      Thermal Decomposition of Anthracite: Part I," *Tanso*, 38, 14
      (1964).

R19. S. Kubata, "Studies in Carbon Powders for Telephone Transmitter-
      Thermal Decomposition of Anthracite: Part II," *Tanso*, 39, 3
      (1964).

R20. L. A. Lyapina, L. A. Lebedev, N. D. Zakharov, and S. V. Orekhov,
      "Study of Changes in the Electric Conductivity of Carbon Black-
      Filled Rubbers During Uniaxial Elongation to Failure," *V. Sh.*,
      Khim. Tekhnol. Ser. Kauchuk. Rezina, 34, 41 (1975).

R21. E. D. Macklen, "Investigation of Electrical Contact Properties
      of Granular Carbon Aggregates," *Brit. J. Appl. Phys.*, 12, 443
      (1961).

R22. E. D. Macklen and B. Hodgson, "Electrical Contact Properties
      of Granular Carbon Aggregates. Part 2: Correlation Between
      Contact Resistance and Oxygen Content," *Brit. J. Appl. Phys.*,
      13, 171 (1962).

R23. E. D. Macklen, "Electrical Contact Properties of Granular
      Carbon Aggregates. Part 3: Effects of Prolonged Air Oxida-
      tion," *Brit. J. Appl. Phys.*, 14, 28 (1963).

R24. E. D. Macklen, "Electrical Contact Properties of Granular
      Carbon Aggregates. Part 4: Investigation of Mechanical
      Ageing," *Brit. J. Appl. Phys.*, 16, 69 (1965).

R25. M. A. Magurpov and K. M. Gafurav, "Electrical Conductivity of
      an Epoxy Resin Filled with Graphite," *Uspekhi Khim. Zh.*, 13,
      54 (1969).

R26. S. Marinkovic, P. W. Whang, A. Nararrete, and P. L. Walker, Jr.,
      "Natural Graphite--CVD Carbon Composites," *Tanso*, 87, 130 (1976).

R27. S. Marinkovic, C. Suznjevic, and M. Djordevic, "Pressure Depend-
      ence of the Electrical Resistivity of Graphite Powder and its
      Mixtures," *Phys. Status Solidi*, A, 4, 743 (1971).

R28. S. Miyanchi, "Electrical Properties of Carbon Black-Graft-
      Polymers Cross-Linked with Peroxide--Divinyl Monomer System,"
      J. Soc. Mater. Sci. Jap., 25, 1005 (1976).

R29. Y. Murase, "Variation of Electrical Resistivity of Graphite
      Powder During Heating in Air," *Nagoya Kogyo Gijutsu Shikensho
      Hokoku*, 22, 116-24 (1973).

R30. H. Nakano and M. Mizuta, "Addition of a New Electroconductive
      Carbon Black-to-Thermoplastics," *Purasuchikkusu*, 26, 67 (1975).

R31. L. K. H. Van Beck and B. I. C. F. van Pul, "Non-Ohmic Behavior
      of Carbon Black-Loaded Rubbers," *Carbon*, 2, 121 (1964).

R32. Yu I. Vasilenok, "Increase in the Electric Conductivity of
      Carbon Black-Filled Polyethylene," *Vysokomol. Soedin.*, A, 15,
      2689-92 (1973).

R33. A. Volt, "Electrical Conductance of Carbon Black," *Rubber Age,*
      95, 746 (1964).

S. Effect of Pressure on Electrical Transport Properties[*]

S1. R. G. Arkhipov, V. V. Kechin, A. I. Likhter, and Yu A. Pospelov,
     "Galvanomagnetic Effects in Graphite and the Deformation of the
     Electron Spectrum of Graphite under Pressure," *J. Exp. Theor.
     Phys.* (USSR), 44, 1964 (1963) [trans. *Soviet Phys. JETP*, 17,
     1321 (1963)].

S2. E. S. Itskevich and L. M. Fisher, "Measurement of the Shubnikov-
     de Haas van Alphen Effect in Graphite at Pressure up to 8 kbar,"
     Zh. Exp. Teor. Fiz. Pis'ma, 5, 141-144 (1967) [trans. *Soviet
     Phys. JETP Lett.*, 5, 114 (1967)].

---

[*]See also C24, C27, R12, R27, W20.

S3. V. V. Kechin, A. I. Likhter, and G. N. Stepanov, "Determination of the Number of Carriers in Pyrolytic Graphite Under Pressure," *Fiz. Tverd. Tela* (USSR), 10, 1242 (1968) [trans. *Soviet Phys. Solid State*, 10, 987 (1968)].

S4. A. I. Likhter and V. V. Kechin, "Investigation of the Dependence of the Galvanomagnetic Effects in Graphite on Temperature and Pressure," *Fiz. Tverd. Tela* (USSR), 5, 3066 (1963) [trans. *Soviet Phys. Solid State*, 10, 987 (1968)].

S5. K. Noto and T. Tsuzuku, "A Note on the Pressure Dependence of Energy Band Parameters of Graphite," *Carbon*, 12, 349 (1974).

S6. K. Noto and T. Tsuzuku, "On the Pressure Dependence of Electrical Conductivity of Graphite," *J. Phys. Soc. Japan*, 35, 1649 (1973).

S7. N. Okuyama, H. Yasunaga, and S. Minomura, "Dependence of the Resistance on Pressure in the c-Direction of Pyrolytic and Natural Graphite," *Jpn. J. Appl. Phys.*, 10, 1645 (1971).

S8. G. A. Samara and H. G. Drickamer, "Effect of Pressure on the Resistivity of Pyrolytic Graphite," *J. Chem. Phys.*, 37, 471 (1962).

S9. H. Sodolski, A. Szumilo, and B. Jachym, "Influence of Pressure on the Electrical Conductivity of Polyester Polymer-Carbon Black Compositions," *Acta. Phys. Pol. Acad.*, A, 51, 217 (1977).

S10. I. L. Spain, "The Effect of Hydrostatic Pressure on the Galvanomagnetic Properties of Graphite," *Electronic Density of States* (L. H. Bennett, ed.), NBS Special Publ. 323, 717 (1971).

S11. I. L. Spain, "The Pressure Dependence of the Conductivity of Graphite," *Carbon*, 14, 229 (1976).

S12. J. Stankiewicz and R. L. White, "Carbon Resistors as Pressure Gauges," *Rev. Sci. Instrum.*, 42, 1067 (1971).

S13. M. L. Yeoman and D. A. Young, "The Anisotropic Pressure Dependence of Conduction in Well-Oriented Pyrolytic Graphite," *J. Phys.* (U.K.), C, 2, 1742 (1969).

T. Transport Properties in Very High Magnetic Fields[*]

T1. N. B. Brandt, G. A. Kapustin, V. G. Karavayer, A. S. Kotosonov, and Ye. A. Svistova, "Investigation of the Galvanomagnetic Properties of Graphite in Magnetic Fields up to 500 koe (50T) at Low Temperatures," *J. Exp. Theor. Phys.* (USSR), 67, 1136 (1974) [trans. *Sov. Phys. JETP*, 40, 564 (1975)].

T2. L. W. Kreps and J. A. Woollam, "Temperature and Field Dependencies of Galvanomagnetic Effects in Graphite," *Carbon*, 15, 403 (1977).

---

[*]See also C12, O4, O11, W23, W25-W27, W29, W33.

T3. W. H. Lowrey and I. L. Spain, "High Field Galvanomagnetic
     Properties of Graphite," *Sol. State Comm.* (U.K.), 22, 615
     (1977).

T4. J. W. McClure and W. J. Spry, "Linear Magnetoresistance in the
     Quantum Limit in Graphite," *Phys. Rev.*, 165(3), 809-815 (1968).

T5. K. Sugihara and J. A. Woollam, "Anomalous Galvanomagnetic Prop-
     erties of Graphite in Strong Magnetic Fields," *J. Phys. Soc.
     Japan,* 45, 1891 (1978).

T6. J. A. Woollam, "Graphite Carrier Locations and Quantum Trans-
     port to 10 T (100 KG)," *Phys. Rev. Solid State,* B, 3, 1148
     (1971).

T7. J. A. Woollam, "Nonlinear Extreme Quantum Limit Resistivity in
     Graphite," *Phys. Lett.*, A, 32, 371-372 (1970).

T8. J. A. Woollam, D. J. Sellmyer, R. O. Dillon, and I. L. Spain,
     "Quantum Limit Studies in Single Crystal and Pyrolytic Graphite,"
     Low Temperature Physics (LT13), Vol. 4 (K. D. Timmerhaus,
     W. J. O'Sullivan, and E. F. Hammel, eds.), Plenum Publ. Corp.,
     New York and London, 1974, p. 358.

U. Conductivity in High Electric Fields and Very Thin Crystals

U1. Y. Fujibayashi, "Electronic Properties of Very Thin Graphite
     Crystals," *J. Phys. Soc. Japan,* 34, 989 (1973).

U2. Y. Fujibayashi and S. Mizushima, "The Residual Resistivity of
     Thin Graphite Crystals," *J. Phys. Soc. Japan,* 34, 281 (1973).

U3. H. J. Goldsmid and J. M. Cossan, "Non-linear Resistance of
     Thermal and Non-thermal Origin in Bismuth and Reheated Pyro-
     lytic Graphite," *Phys. Lett.*, 8, 221 (1964).

U4. H. Hayashi and K. Yazawa, "Multiple Kink Effect in Graphite,"
     J. Phys. Soc. Japan, 32, 1545 (1972).

U5. S. Mizushima and T. Endo, "Observation of the Esaki Kink Effect
     in Graphite," *J. Phys. Soc. Japan,* 24, 1402 (1968).

U6. S. Mizushima, Y. Fujibayashi, and K. Shiki, "Electric Resis-
     tivity and Hall Coefficient of Very Thin Graphite Crystals,"
     J. Phys. Soc. Japan, 30, 299 (1971).

U7. S. Mizushima and Y. Fujibayashi, "Esaki Kink Effect in Graphite,"
     J. Phys. Soc. Japan, 38, 1027 (1975).

U8. B. Robrieux, R. Faure, and J. P. Dussaulcy, "Resistivity and
     Work Function of Carbon in Very Thin Layers" (in French),
     Comp. Rendu (Paris), B, 278, 659 (1974).

U9. S. Yugo, J. Hayashi, and K. Yazawa, "Multiple Kink Effect in
     Graphite," *Appl. Phys. Lett.*, 17, 339 (1970).

V. Carbon Resistors[*]

V1.   J. Basacoma, "Quality Control on Carbon Film Resistors," *New Electron* (U.K.), 9, 85 (1976).

V2.   T. Ciborowski and K. Goratowski, "Fabrication of Carbon Film Resistors," *Pol. Tech. Rev.* (Poland), 9, 2 (1977).

V3.   B. Day, "Carbon Potentiometers, II," *Design Electronics* (U.K.), 6, 18 (1969).

V4.   K. Hattori, "Carbon Composition Resistors--Miniature, Durable, Practically Immune to External Influence," *JEE*, 76, 67 (March 1973).

V5.   B. K. Jones, "1/F and 1/ΔF Noise Produced by a Radio-frequency Current in a Carbon Resistor," *Electron Lett.* (U.K.), 12, 110 (1976).

V6.   S. Kondo, "Pyrolytic Carbon Film Resistors on Forsterite Ceramic Substrate," *Rev. Elec. Commun. Lab.*, 17, 689 (1969).

V7.   J. Lee and J. S. Kim, "Consideration of Noise Characteristic for Carbon Type Resistors," Col. Thesis Civil Aviat. College, Korea, 11, 97 (1970); *Kor. Sci. Abst.*, 3, 47 (1971).

V8.   H. Liess and A. Sacher, "Voltage Coefficient, Distortion and Noise in Carbon Resistors," *Radio Mentor Electron.*, 34, 551 (1968).

V9.   H. L. Lo, "Relation Between Quality and Current Noise in Carbon Film Resistors," *J. Kor. Inst. Electron. Eng.*, 9, 34 (1972).

V10. C. Lockert, "Thick Film Resistors for Kilovolt Circuits," *Mach. Des.*, 49, 156 (1977).

V11. S. Marsh, "Carbon Film Resistors--The Growth Product," *New Electron.* (U.K.), 8, 62 (1975).

V12. W. McCormick and S. I. Tan, "Vitreous, High Temperature, Film Resistor Material," *IBM Tech. Disclosure Bull.*, 20, 785 (1977).

V13. S. Nakazawa, T. Harada, T. Shimada, and H. Shiomi, "Life Characteristic of Carbon Film Resistors Under Pulsive and DC Operation," *Trans. Inst. Electron. and Commun. Eng. Japan E* (Japan), 59, 35 (1976).

V14. E. Polturak, M. Rappaport, and R. Rosenbaum, "Improved Thermal Contact to Unmodified Carbon Resistors," *Cryogenics* (U.K.), 18, 27 (1978).

V15. K. W. Stanley, "Carbon Film Resistors," *Electron.* (U.K.), 99, 31 (1976).

V16. K. W. Stanley, "Reliability and Stability of Carbon Film Resistors," *Microelectron. Reliab.*, 10, 359 (1971).

---

[*]See also L11, Section W.

V17. K. W. Stanley, "Reliability and Stability of Carbon Film
      Resistors," *Mullard Tech. Commun.*, 11, 223 (1971).

V18. S. Strijobs, "The Manufacture of Carbon-film Resistors Employ-
      ing Fluidized-bed Technology," *Philips, Res. Rep.*, 27, 186
      (1972).

V19. Tien-Shou Wu, T. Shiramatsu; Shen-Li Fu, and Mong-Song Liang,
      "Electrical Properties of Carbon Black/Resin Thick Film Resis-
      tors," *Int. Conf. Thin- Thick-Film, Ntg-Fachber* (Germany), 60,
      34 (1977).

V20. T. Yanagisawa and H. Shiomi, "Degradation of Carbon Film Resis-
      tor in Damp-Heat Environments," *Bull. Electrotech. Lab.* (Japan),
      42, 333 (1978).

V21. T. Yanagisawa and H. Shiomi, "Life and Degradation Characteris-
      tics of Carbon Film Resistor Under On-Off Voltage Stress Opera-
      tion. I," *Bull. Electrotech. Lab.* (Japan), 42, 435 (1978).

W. Carbon Resistance Thermometers, Bolometers, and Transducers[*]

W1. H. Alms, R. Tillmanns, and S. Roth, "Magnetic-Field-Induced
     Temperature Error of Some Low-Temperature Thermometers," *J.
     Phys.* (U.K.), E, 12, 62 (1979).

W2. A. C. Anderson, J. H. Anderson, and M. P. Zaitlin, "Some Obser-
     vations on Resistance Thermometry Below 1 K," *Rev. Sci. Instr.*,
     47, 407 (1976).

W3. R. J. Balcombe, D. J. Emerson, and R. S. Potton, "A Calibration
     Equation for Carbon Resistance Thermometers," *J. Phys., Sci.,
     Instr.*, E., 3, 43 (1970).

W4. W. C. Black, Jr., W. R. Roach, and J. C. Wheatley, "Speer
     Carbon Resistors as Thermometers for Use Below 1 K," *Rev. Sci.
     Instr.*, 35, 587 (1964).

W5. B. L. Booth and A. W. Ewald, "Fabrication of Extremely Small
     Carbon Resistor Thermometers," *Rev. Sci. Inst.*, 40, 1354 (1969).

W6. S. Corsi, G. Dall'Oglio, G. Fantoni, and F. Melchiorri, "Recent
     Advances in Carbon Bolometers as far Infrared Detectors,"
     Infrared Phys., 13, 253-273 (1973).

W7. S. Cumsolo, M. Santini, and M. Vicentini-Missoni, "Interpolation
     and Extrapolation of a Carbon Resistance Thermometer Calibration
     Data in the Liquid Helium II Region," *Cryogenics*, 5, 168 (1965).

W8. D. L. Decker and H. L. Laquer, "Magnetoresistance of Carbon and
     Germanium Thermometers to 60kG," *Cryogenics*, 9, 481 (1969).

---

*See also A4, S12.

W9.   B. Dodson, T. Low, and J. Mochel, "Low-Temperature Thin Graphite Film Thermometers," *Rev. Sci. Instr.*, 48, 290 (1977).

W10. A. Dupre, A. Ian Tterbeek, L. Michiels, and L. Van Neste, "The Use of Graphite Thermometers in Heat Conductivity Experiments Below 1 K," *Cryogenics*, 4, 354 (1964).

W11. R. Gerber, F. Vilim, and K. Zaveta, "Low Temperature Measurements with Carbon Thermometer," *Cesk. Casopis. Fys.*, 15, 340 (1965).

W12. J. H. Greenwood, S. Lebeda, and J. Bernasconi, "The Anisotropic Electrical Resistivity of a Carbon-Fiber Reinforced Plastic Disc and its Use as a Transducer," *J. Phys.*, 8, E, 369 (1975).

W13. A. Greser, et al., "On Measuring Low Temperatures with a Carbon Resister," *Ceske, Casopis, Fys.*, 20, 26 (1970).

W14. I. N. Kalinkina, "Temperature Dependence of the Resistance of Carbon Thermometers," *Cryogenics*, 4, 327 (1964).

W15. C. S. Kang and J. Vanderkooy, "Carbon Resistance Bolometers and Thermometers and Their Characteristics in Field Modulation Studies," *Cryogenics*, 16, 713 (1976).

W16. S. Kobayshi, M. Shinohara and K. Ono, "Thermometry Using 1/8 w Carbon Resistors in a Temperature Region around 10 mK," *Cryogenics*, 16, 597 (1976).

W17. F. J. Kopp and T. Ashworth, "Carbon Resistors as Low Temperature Thermometers," *Rev. Sci. Instru.*, 43, 327-333 (1972).

W18. B. Leontic and J. Lukatela, "Manufacture and Performance of a Film-Type Carbon Resistance Thermometer," *Elektrotehnika Zagreb* (Yugoslavia), 19, 175 (1976).

W19. C. W. Marshall and P. R. Held, "Measurement of Long-term Stability with Electrical Resistance Strain Gauges," *Strain*, 13, 13 (1977).

W20. C. E. Miller, J. W. Dean, and T. M. Flynn, "Commercial Carbon Composition Resistors as Pressure Transducers," *Rev. Sci. Instr.*, 36, 231 (1965).

W21. K. Mineo, "Carbon and Germanium Resistance Thermometers," *Oyo Butsuri*, 38, 155 (1969).

W22. H. Morisaki and K. Yazawa, "Memory-Switching Phenomena in Amorphous Carbon," *Rep. Univ. Electro-Commun.* (Japan), 25, 255 (1975).

W23. L. J. Nearinger and Y. Shapira, "Low Temperature Thermometry in High Magnetic Fields--I. Carbon Resistors," *Rev. Sci. Instr.*, 40, 1314 (1969).

W24. K. Noto, N. Kobayashi, and Y. Muto, "High Sensitivity Temperature Sensors at Low Temperatures," *J. Appl. Phys.* (Japan), 15, 2449 (1976).

W25. S. Saito and T. Sato, "Matsushita Carbon Resistors as Ther-
      mometers for Use at Low Temperatures and in High Fields," *Rev.
      Sci. Instrum.*, 46, 1226 (1975).

W26. S. Saito, T. Sato, and M. Krusius (eds.), "Low Temperature
      Thermometry in Magnetic Fields with Matsushita Carbon Resistors,"
      in *Proceedings of the 14th International Conference on Low
      Temperature Physics, Otaniemi, Finland, Aug. 14-20, 1975, Pt.
      IV,* North-Holland, Amsterdam (Netherlands), 1975.

W27. H. H. Sample and L. G. Rubin, "Instrumentation and Methods for
      Low Temperature Measurements in High Magnetic Fields," *Cryo-
      genics* (U.K.), 17, 597 (1977).

W28. J. Sanchez, A. Benoit, and J. Flouquet, "Speer Carbon Resistance
      Thermometer Magnetoresistance Effect," *Rev. Sci. Instru.*, 48,
      1090 (1977).

W29. W. F. Schlosser and R. H. Mannings, "A Method for Reducing the
      Effective Magnetoresistance of Carbon Resistor Thermometer,"
      Cryogenics, 12, 225 (1972).

W30. E. H. Schulte, "Carbon Resistors for Cryogenic Temperature
      Measurement," *Cryogenics*, 6, 321 (1966).

W31. S. F. Strait and B. Bartman, "Thermal Boundry Resistance
      Difference Between Carbon Resistors in Contact with Helium
      Liquid and Vapour," *Cryogenics*, 9, 328 (1969).

W32. J. M. Swartz, C. F. Clark, D. A. Johns, and D. L. Swartz,
      "Germanium and Carbon Glass Resistance Thermometry: A Compari-
      son of Characteristics, Stability and Construction," in *Pro-
      ceedings of the Sixth International Cryogenic Engineering Con-
      ference, Grenoble, May 1976* (F. R. S. Mendelssohn, ed.), IPC
      SCI and Technology Press, Guildford (England), 1976, p. 198.

W33. J. M. Swartz, J. R. Gaines, L. G. Rubin, "Magnetoresistance of
      Carbon-Glass," *Rev. Sci. Instru.*, 46, 1177 (1975).

W34. J. R. Thompson and J. O. Thomson, "Low-Temperature Magneto-
      resistance of a Speer Carbon Resistance Thermometer," *Rev.
      Sci. Instru.*, 48, 1713 (1977).

W35. E. A. Tishchenko, "Cooked Carbon Bolometer," *Prib. Tekh. Eksp.*,
      5, 215 (1970).

W36. H. Van Dael and W. M. Star, "Manufacture of a Film-Type Carbon
      Resistance Thermometer," *Phys. Soc. Proc. (Solid State Phys.),*
      E, 2, 910 (1969).

W37. S. F. Vorfolemeev, L. A. Pekal'n, B. I. Al'shin, G. S. Avilov,
      and L. B. Belyanskii, "Wide Range Carbon Thermometers for Low
      Temperatures," *Prib. Tekh. Eksp.* (USSR), 20, 262 (1977).

W38. D. A. Ward, "The Interpolation of Germanium and Carbon Cryo-
      genic Thermometer Calibrations," *Cryogenics*, 12, 209 (1972).

W39. G. Wehr, G. Sieber, and K. Boning, "Carbon Resistors as Low
      Temperature Sensors in Low Temperature Reactor Irradiation
      Experiments," *Cryogenics*, 17, 43 (1977).

X. Carbon and Composite Electrodes,
Brushes, Contacts, Etc.[*]

X1. Y. Asai and Y. Nagano, "Effect of Manufacturing Conditions on Carbon Brush Characteristics," *Tanso*, 45, 6 (1966).

X2. J. J. Bates, "Carbon Fiber Brushes for Electrical Machines," in *Carbon Fibers in Engineering* (M. Langley, ed.), McGraw-Hill, London, 1973, p. 134.

X3. J. J. Bates and R. Powell, "Relative Current-Switching Properties of Carbon-Fibre Brushes and Solid Brushes and the Implications for Commutator Machines," *Proc. Inst. Elec. Eng.*, 118, 604 (1971).

X4. A. L. Beilby and B. R. Mather, "Resistance Effects of Two Types of Carbon Paste Electrodes," *Ancel. Chem.*, 37, 766 (1965).

X5. B. A. Borisov, E. I. Ivanova, and I. G. Mel'nikova, "Properties of the Graphites Used as Heaters in the Manufacture of Fused Silica," *Opt.-Mekh. Proml.* (USSR), 43, 60 (1976) [trans. *Soviet J. Opt. Technol.* (U.S.), 43, 184 (1976)].

X6. D. L. Dixon, "Development with Carbons for Current Collection," *R. Eng. J.*, 2, 46 (Sept. 1973).

X7. N. S. Dyadenko, A. I. Lun'ko, and T. V. Kholoptseva, "Electrical Resistivity of Copper Graphite Powder Composites (Contact Materials)," *Porosh. Metal.* (USSR), 12, 62 (1973) [trans. *Sov. Powder Metall. and Met. Cer.* (U.S.), 12, 974 (1973)].

X8. C. Foley, "Electrical Carbon and the Challenge of Railways," *Railway Eng.*, 4, 112 (1975).

X9. R. Fuchs and W. Herzog, "Semiconducting Composite Materials for Electrical Machine, Transformer and Cable Insulation," *Elektrotech. Z. Etz* (Germany), B, 30, 390 (1978).

X10. W. Graham and E. A. Harvey, "The Electrical Resistance of Fluidized Beds of Coke and Graphite," *Can. J. Chem. Eng.*, 43, 146 (1965).

X11. W. Hanfe and H. Schreiner, "Properties of Silver-Graphite Sintered Contact Materials for Power Applications," *Bull. Assoc. Suisse. Electr.* (Switzerland), 67, 1340 (1976).

X12. T. Hirayama, "Silver-Graphite-Nickel Oxide Electrical Contact Materials," *Japan Kokai*, 75, 104 (1974).

X13. K. Ilto and R. Ano, "Current Distribution Characteristics of Parallel Carbon Brushes," *Tanso (Carbons)*, 35, 8 (1963).

X14. J. L. Johnson and L. E. Moberly, "High-Current Brushes. I. Effect of Brush and Ring Materials," in Proceedings of the 23rd Annual Holm Conference on Electrical Contacts, *IEEE Trans. Components, Hybrids and Manuf. Technol. (CHMT)*, 1, 36 (1978).

---

[*]See also R7-R15, R25, R26.

X15. K. Kakubu, "Brush Wear and Commutation Film," *Tanso*, 39, 9 (1964).

X16. T. Kitagawa, H. Takao, and Y. Feyikawa, "Fundamental Studies of Carbon Paste Electrode," *Japan Analyst*, 15, 446 (1966).

X17. J. T. Kung and M. P. Amason, "Electrical Conductive Character- istics of Graphite Composite Structures," *IEEE*, XI, 403 (1977).

X18. A. M. Kuz'min, V. I. Volga, E. M. Izotova, and L. M. Terent'eva, "Investigation into Electrical Properties of a Copper Carbon Fibre Composite Material at Cryogenic and High Temperatures," *Fiz. Khim. Obrab. Mater.* (USSR), 3, 132 (1977).

X19. J. Lindquist, "A Study of Seven Different Carbon Paste Elec- trodes," *J. Electroanal. Chem. Interfacial Electrochem.*, 52, 37 (1974).

X20. V. Mooney, "The Use of Pure Carbon for Permanent Percutaneous Electrical Connector Systems," Report N74-32016/9GA, Univ. of Southern Calif.; Rehabilitation Eng. Center, 1970.

X21. J. Orr, "Pressure and High Current Effects on a Graphite Con- tact," *J. Phys. Chem. Solids*, 24, 1695 (1963).

X22. E. C. Pike and J. E. Thompson, "On the Possible Application of Pyrolytic Graphite as a Brush Material and a Critical Assess- ment of the Use of Brushes with Truncated Conicial Tips to Study the Running Characteristics of Graphite Materials," *Wear*, 13, 247 (1969).

X23. B. Prakash and B. Marincek, "Electrical Conductivity of Coke and Iron Ore Mixtures at High Temperatures," *Trans. Indian Inst. Metals*, 22, 10 (1969).

X24. G. E. Ross, "Carbon as an Electrical Engineering Material," *Focus*, 159, 16 (1971).

X25. L. A. Scruggs and W. J. Gajda, Jr., "Low Frequency Conductivity of Unidirectional Graphite/Epoxy Composite Samples," *IEEE*, XI, 396 (1977).

X26. M. Szeifert, "Variation of Contact Resistance at the Cable Junc- tion of Carbon Brushes," *Villamossag* (Hungary), 26, 69 (1978).

X27. L. E. Vrublevskii, B. N. Vinogradov, V. S. Gorshkov, L. N. Repyakh, and T. A. Khmelevskaya, "Study of the Phase Composition and Structure of Current-Conducting Carbon-Cement Stone Compos- ites," *Izv. Sib. Otd. Akad. Nauk SSR, Ser. Tekh. Nauk Izv.*, 3, 123 (1976).

X28. W. F. Walker and R. E. Heintz, "Conductivity Measurements of Graphite/Epoxy Composite Laminates at UHF Frequencies," *IEEE*, XI, 410 (1977).

X29. D. S. Yas, "Copper Graphite Materials with Additives of Graphite Granules Plated with Copper. Antifriction Properties of Copper Graphite Materials Containing Different Amounts of Graphite Granules Plated with Copper," *Porosh. Met.*, 13, 17-21 (1973).

X30. D. S. Yas, "Metal Graphite Materials of High Graphite Content
      and Some Methods of Preparation," *Sov. Powder Metall. Met. Cer.*,
      15, 27 (1976).

Y. Techniques for Measurement

Y1.   J. D. Crowley and T. A. Rabson, "Contactless Method of Measuring
      Resistivity," *Rev. Sci. Instr.*, 47, 712 (1976).

Y2.   I. Drummond and A. R. Ubbelohde, "Measurements Using Large Elec-
      trical Pulses," *J. Phys.* (U.K.), E, 10, 1114 (1977).

Y3.   W. C. Dunlap, "Conductivity Measurements on Solids," in *Methods
      of Experimental Physics,* Vol. 6 B (K. Lark-Horovitz and V. A.
      Johnson, eds.), Academic, New York and London, 1959.

Y4.   S. Flandrois, "Effect Hall d' Echatillons de Forme Quelquonque:
      Application aux Poudres," *J. Chim. Phys.*, 66, 444 (1969).

Y5.   H. Fritzsche, "The Hall Effect," in *Methods of Experimental
      Physics,* Vol. 6 B (K. Lark-Horovitz and V. A. Johnson, eds.),
      Academic, New York and London, 1959.

Y6.   V. V. Gal'perm, "Apparatus for Resistivity Measurements on Coke
      Briquettes," *Coke Chem.* (USSR), 4, 19 (1977).

Y7.   S. M. Horszowski, "A Pulse Method for Measuring the Electrical
      Resistance of Semiconducting Diamond," *J. Sci. Instr.*, 44, 296
      (1967).

Y8.   H. L. Libby, *Introduction to Electromagnetic Non-Destructive
      Test Methods,* Wiley-Intersci., New York, 1977.

Y9.   H. C. Montgomery, "Method for Measuring Electrical Resistivity
      of Anisotropic Materials," *J. Appl. Phys.* (U.S.), 42, 2971
      (1971).

Y10. C. N. Owston, "Eddy Current Methods for the Examination of
      Carbon Fibre Reinforced Epoxy Resins," *Mater. Eval.*, 34, 237
      (1976).

Y11. G. L. Pearson, "Magnetoresistance," in *Methods of Experimental
      Physics,* Vol. 6 B (K. Lark-Horovitz and V. A. Johnson, eds.),
      Academic, New York and London, 1959.

Y12. R. D. Swenmunson, U. Even, and J. C. Thompson, "Double Induction
      Method for Electrodeless Determination of Hall Mobility," *Rev.
      Sci. Instr.*, 49, 519 (1978).

Y13. L. J. Van der Pauw, "A Method of Measuring Specific Resistivity
      and Hall Effect of Discs of Arbitrary Shape," *Phillips Res.
      Rep.*, 13, 1 (1958).

Y14. C. Zeller, A. Denenstein, and G. M. T. Foley, "Contactless
      Technique for Measurement of Electrical Resistivity of Aniso-
      tropic Material," *Rev. Sci. Instr.*, 50, 602 (1979).

Z. Recent Publications

ZD1.   A. A. Bright, "Negative Magnetoresistance in Pregraphite
        Carbons," *Phys. Rev.*, B20, 5142 (1980).

ZD2.   S. Mizushima, "Various Aspects of Interaction Between Electrons
        and Phonons in Graphite," *Carbon*, 17, 187 (1979).

ZE1.   K. Sugihara, K. Kawamura, and T. Tsuzuku, "Temperature Depend-
        ence of the Average Mobility of Graphite," *J. Phys. Soc. Japan*,
        47, 1210 (1979).

ZF1.   C. Ayache, "Analysis of Transport Properties of Graphite at
        Low Temperatures," *Physica*, B99, 509 (1980).

ZF2.   I. L. Spain, "Galvanomagnetic Effects in Graphite," *Carbon*,
        17, 209 (1979).

ZH1.   Y. Hishiyama, Y. Kaburagi, and A. Ono, "Variable Range Hopping
        Conduction and Negative Magnetoresistance of Disordered Carbons
        at Low Temperature," *Carbon*, 17, 265 (1979).

ZH2.   S. Mrozowski, "Specific Heat Anomalies and Spin-Spin Inter-
        actions in Carbons: A Review," *J. Low Temp. Phys.*, 35, 231
        (1979).

ZJ1.   K. Rozwadowska, "Preparation and DC Conductivity of Low Tempera-
        ture Glassy Carbon," *Phys. Status Solidi A: Appl. Res.*, 54, 93
        (1979).

ZK1.   C. N. Owston, "Electrical Properties of Single Carbon Fibers,"
        J. Phys. D: Appl. Phys., 13, 1615 (1970).

ZK2.   F. L. Vogel and J. Gan, "High Electrical Conductivity of Inter-
        calated Graphite Composites," *Met. Soc. AIME*, 1978, 166.

ZM1.   A. A. Ovchinnikov and A. A. Belyi, "Density of States and Posi-
        tion of Fermi Level in Doped Graphite," *Sov. Phys. Semicond.*,
        8, 462 (1974).

ZN1.   G. Micaud, J. Rappenau, A. Pacault, A. Marchand, and J. Amiell,
        "Processus de Graphitization sou l'Effet des Rayonnements Neu-
        troniques," *J. Chim. Phys.*, 1969, 129.

ZO1.   T. Tsuzuku, "Anisotropic Electrical Conduction in Relation to
        Stacking Disorder in Graphite," *Carbon*, 17, 293 (1979).

ZQ1.   F. Batallan, I. Rosenman, C. Simon, G. Furdin, and H. Fuzellier,
        "The Electronic Structure of the Graphite Acceptor Compounds,"
        Physica, B99, 411 (1980).

ZQ2.   W. Eberhardt, I. T. McGovern, E. W. Plummer, and J. E. Fischer,
        "Charge Transfer and Non-Rigid Band Effects in the Graphite
        Compound LiC_6," *Phys. Rev. Lett.*, 44, 200 (1980).

ZQ3.   M. Endo, K. Mori, T. Koyama, and M. Inagaki, "Preparation and
        Electrical Properties of Graphite Fiber Nitrate," *Denki Gakkai
        Ronbunsyu*, 98, 249 (1978).

ZQ4. J. E. Fischer, "Graphite Intercalation Compounds: Electronic Properties and their Correlation with Chemistry," *Physica*, B99, 383 (1980).

ZQ5. N. A. W. Holzworth, S. Rabii, and L. A. Girifalco, "Theoretical Study of Lithium Graphite I Band Structure, Density of States, Fermi Surface Properties," *Phys. Rev. B: Solid State*, 18, 5190 (1978).

ZQ6. N. A. W. Holzworth, S. Rabii, and L. A. Girifalco, "Theoretical Study of Lithium Graphite II Spatial Distribution of Valence Electrons," *Phys. Rev. B: Solid State*, 18, 5206 (1978).

ZQ7. H. Kamimura, K. Nakao, T. Ohno, and T. Inoshita, "Electronic Structure and Properties of Alkali-Graphite Intercalation Compounds," *Physica*, B99, 401 (1980).

ZQ8. Y. Koike, H. Suematsu, K. Higuchi, and S. Tanuma, "Superconductivity in Graphite-Alkali Metal Intercalation Compounds," *Physica*, B99, 503 (1980).

ZQ9. E. McCrae, D. Billaud, J. F. Marêché, and A. Hérold, "Basal Plane Resistivity of Alkali-Metal-Graphite Compounds," *Physica*, B99, 489 (1980).

ZQ10. I. T. McGovern, W. Eberhardt, E. W. Plummer, and J. E. Fischer, "The Band Structures of Graphite and Graphite Intercalation Compounds as Determined by Angle Resolved Photoemission Using Synchrotron Radiation," *Physica*, B99, 415 (1980).

ZQ11. E. McCrae, A. Metrot, P. Willmann, and A. Hérold, "Evolution of Electrictrical Resistivity for Graphite-H_2SO_4 and Graphite-Sulphuric Oleum Compounds," *Physica*, B99, 541 (1980).

ZQ12. A. Metrot, P. Willmann, T. McRae, and A. Hérold, "Remarques sur les Variations de la Resistivité Electrique d'un Pyrographite au Cours de l'Insertion Electrochemiques de H_2SO_4," *Carbon*, 17, 182 (1979).

ZQ13. T. Ohno, K. Nakao and H. Kamimura, "Self Consistent Band Calculation of the Band Structure of C_BK Including the Charge Transfer Effect," *J. Phys. Soc. Japan*, 47, 1125 (1979).

ZQ14. D. G. Onn, G. M. T. Foley, and J. E. Fischer, "Electronic Properties, Resistive Anomalies, and Phase Transitions in the Graphite Intercalation Compounds with K, Rb, Cs," *Phys. Rev.*, B19, 6474 (1979).

ZQ15. P. Pfluger, P. Oelhafer, H. U. Kunz, R. Jeber, E. Hauser, K. P. Ackermann, M. Muller, and H. J. Guntherodt, "Electronic Properties of Graphite Intercalation Compounds," *Physica*, B99, 395 (1980).

ZQ16. L. Pietronero, S. Strässler, H. R. Zeller, and M. J. Rice, "Theory of the Electrical Conductivity of Acceptor Compounds of Graphite," *Physica*, B99, 499 (1980).

ZQ17. H. A. Resing, F. L. Vogel, and T. C. Wu, "Carrier Density from ^{19}F Nuclear Magnetic Resonance Line Shapes for Graphite Intercalated with SbF_5," *Mat. Sci. Eng.*, 41, 113 (1979).

ZQ18. J. J. Ritsko and E. J. Mele, "The Electronic Structure of Ferric Chloride Intercalated Graphite," *Physica*, B99, 425 (1980).

ZQ19. M. Sano and H. Inokuchi, "The Conductivity of Graphite-Alkali Metal-Hydrogen Ternary Systems," *Chem. Lett.*, 1979, 405.

ZQ20. S. C. Singhal, "Electrical Conductivity of Copper/SbF$_5$-Intercalated Graphite Composite Wires," *Physica*, B99, 536 (1980).

ZQ21. L. Streifinger, H. P. Boem, R. Schlogl, and R. Pentrider, "The Electrical Resistivity of Compacts of Graphite Intercalation Compounds with SbF$_5$ and SbCl$_5$," *Carbon*, 17, 195 (1979).

ZQ22. H. Suematsu, S. Tanuma, and K. Higuchi, "Electronic Properties of Graphite Acceptor Compounds," *Physica*, B99, 420 (1980).

ZQ23. S. Tanuma, K. Higuchi, H. Suematsu, and Y. Koike, "Fermi Surface, Transport and Superconducting Properties of Graphite Intercalation Compounds," *J. Phys.*, C8, 1104 (1978).

ZQ24. D. P. Di Vincenzo, N. A. W. Holzwarth, and S. Rabii, "The Electronic Structure of KC$_8$," *Physica*, B99, 406 (1980).

ZQ25. F. L. Vogel, H. Fuzellier, C. Zeller, and E. J. McRae, "In-plane Electrical Resistivity of Nitric Acid Intercalated Graphite," *Carbon*, 17, 255 (1979).

ZQ26. F. L. Vogel and C. Zeller, "High Electrical Conductivity Graphite Intercalation Compounds," *Mol. Met.*, 1979, 289.

ZT1. K. Sugihara, "Anomalous Galvanomagnetic Properties of Graphite in the Quantum Limit," *Carbon*, 17, 201 (1979).

ZV1. S. Nakazawa, "Life Characteristics of Carbon Film Resistors Under Pulsive and DC Operation," *Trans. Inst. Electron. & Comm. Eng. Japan,* 59, 35 (1976).

ZX1. J. P. Decruppe, F. Dujardin, M. F. Charlier, A. Charlier, "Etude de la Resistivité du coups Pulverulents à Base de Carbon en Fonction de la Compression," *Carbon*, 17, 237 (1979).

ZX2. J. R. Lakin, "Assessment Techniques for Graphite Electrodes," *Fuel*, 57, 151 (1978).

ZX3. R. E. Sioda, "Limiting Current at Porous Graphite Electrodes Under Flow Conditions," *J. Appl. Electrochem.*, 5, 221 (1975).

ZX4. P. G. Tanner, "Using Carbon Fiber Brushesto Improve Commutation," *Electron. Times*, 4320, 7 (1975).

ZY1. E. J. McCrae, J. F. Mareche, and A. Herold, "Contactless Resistivity Measurements: A Technique Adapted to Graphite Intercalation Compounds," *J. Phys. E: Sci. Instr.*, 13, 241 (1980).

Many interesting papers on intercalation compounds of graphite will appear shortly in *Synthetic Metals*, Vol. 2, which includes papers from the 2nd Int. Conf. Int. Compounds of Graphite, Provincetown (F. L. Vogel, ed.).

AUTHOR INDEX

Numbers in parentheses are reference numbers and indicate that an author's work is referred to although the name is not cited in the text. Underlined numbers give the page on which the complete reference is listed.